Springer Theses

Recognizing Outstanding Ph.D. Research

Aims and Scope

The series "Springer Theses" brings together a selection of the very best Ph.D. theses from around the world and across the physical sciences. Nominated and endorsed by two recognized specialists, each published volume has been selected for its scientific excellence and the high impact of its contents for the pertinent field of research. For greater accessibility to non-specialists, the published versions include an extended introduction, as well as a foreword by the student's supervisor explaining the special relevance of the work for the field. As a whole, the series will provide a valuable resource both for newcomers to the research fields described, and for other scientists seeking detailed background information on special questions. Finally, it provides an accredited documentation of the valuable contributions made by today's younger generation of scientists.

Theses may be nominated for publication in this series by heads of department at internationally leading universities or institutes and should fulfill all of the following criteria

- They must be written in good English.
- The topic should fall within the confines of Chemistry, Physics, Earth Sciences, Engineering and related interdisciplinary fields such as Materials, Nanoscience, Chemical Engineering, Complex Systems and Biophysics.
- The work reported in the thesis must represent a significant scientific advance.
- If the thesis includes previously published material, permission to reproduce this must be gained from the respective copyright holder (a maximum 30% of the thesis should be a verbatim reproduction from the author's previous publications).
- They must have been examined and passed during the 12 months prior to nomination.
- Each thesis should include a foreword by the supervisor outlining the significance of its content.
- The theses should have a clearly defined structure including an introduction accessible to new PhD students and scientists not expert in the relevant field.

Indexed by zbMATH.

More information about this series at http://www.springer.com/series/8790

Kieran Finn

Geometric Approaches to Quantum Field Theory

Doctoral Thesis accepted by
University of Manchester, Manchester, United Kingdom

Springer

Author
Dr. Kieran Finn
Department of Physics
and Astronomy
University of Manchester
Manchester, UK

Supervisor
Prof. Apostolos Pilaftsis
University of Manchester
Manchester, UK

ISSN 2190-5053 ISSN 2190-5061 (electronic)
Springer Theses
ISBN 978-3-030-85268-9 ISBN 978-3-030-85269-6 (eBook)
https://doi.org/10.1007/978-3-030-85269-6

This Springer imprint is published by the registered company Springer Nature Switzerland AG
The registered company address is: Gewerbestrasse 11, 6330 Cham, Switzerland

Dedicated to my grandmother, Peggy Anne Green (24th May 1939–23rd March 2020)

Supervisor's Foreword

Since antiquity, humans have been fascinated to understand better the origin of the Cosmos and, in particular, the way this is imprinted in the constellations of stars, the motion of planets, and other celestial objects. Back in the sixth century BC, Pythagoreans believed that the divine origin of the Cosmos could be understood geometrically by describing the distance from the Earth of the five known moving stars (planets) as radii of regular solids (more often known as Platonic solids) all inscribable in three-dimensional spheres. Refinement of such approaches led later to the development of a Ptolemaic geocentric system of deferents and epicycles, as well as to alternative, but less influential, schools of thought initiated by Aristarchus and Seleucus who proposed a heliocentric model around the third century BC.

In the sixteenth century AD, Tycho Brahe's measurements of planetary motion, Galileo's observations of the phases of Venus, as well as the works of Copernicus and Kepler, strongly disfavoured geocentric models, and the heliocentric system saw wider acceptance. However, the discrepancies between the two models went unresolved until the seventeenth century AD with the formulation of Newton's law of universal attraction. The full consequences of Newton's frame-dependent dynamics were eliminated only through the introduction of fictitious forces by d'Alembert and Coriolis in the eighteenth century AD. Our modern view is that both Tycho's geocentric and heliocentric models are fully equivalent provided the appropriate fictitious forces are considered and are related by means of a coordinate transformation.

While inertial frames might appear more physically intuitive than non-inertial frames, there should be a priori no reason to prefer one system over the other, apart from convenience or ease in calculations. In his nominal theory of General Relativity in the twentieth century AD, Einstein addressed successfully Newton's frame-dependent dynamics by introducing a fully differential-geometric description of the Cosmos. Any form of energy or matter will give rise to a curved manifold upon which geodesics become the only allowed trajectory of motion for any object or elementary particle, including light itself. Einstein's pioneering ideas revolutionised our understanding of fundamental physics, as they imply that true laws of nature should be frame covariant, and so manifestly independent of our description.

With the advent of Quantum Mechanics and Quantum Field Theory in the last century, the issue of quantising Einstein's theory of gravitation has resulted in perhaps the most famous long-standing problem of the twenty-first century. Specifically, finding a UV complete theory of Quantum Gravity constitutes a huge milestone in the advancement of fundamental physics. Thus far, several approaches have been put forward in the literature, such as loop quantum gravity, superstring, and membrane theories, but all these proved to have limited success.

Motivated by Einstein's theory of Gravitation, Vilkovisky and DeWitt (VDW) in the early 1980s proposed a covariant differential-geometric approach to Quantum Gravity, which includes not only the coordinate space but an entire field space. This was an attractive idea as covariant field-space actions may resolve the problems of gauge dependence in gauge quantum field theories.

The VDW effective action may solve a number of problems that would require knowledge of the off-shell dynamics of a quantum system, which are highly frame-dependent in the configuration space. In particular, problems of great interest and importance are as follows: (i) the gauge-independent definition of effective charges in non-Abelian gauge theories; (ii) the gauge-invariant description of unstable particle dynamics within the context of S-matrix theory; (iii) the proper field-reparameterization invariant definition of electroweak precision observables, including Veltman's electroweak parameter; (iv) the unique field-reparameterization invariant expansions of effective field theories such as Standard Model Effective Field Theory (SMEFT); and (v) a frame covariant description of cosmological inflation in models with multiple scalar fields that may act as inflatons.

One pressing question regarding the VDW effective action was its uniqueness. This issue of uniqueness has been the topic of numerous studies. In the same context, it was recently realised that in addition to the classical action, the path-integral measure responsible for quantizing the theory using Feynman's approach to QFT can also introduce frame dependence through the gravity sector, thereby ruining the central mission of the VDW formalism. Fortunately, as it will be shown in this thesis, this ambiguity can be eliminated by properly defining the model functions of the action and by identifying a novel one that governs the path-integral measure.

Another previously unsurmountable difficulty of the VDW formalism was the consistent inclusion of fermions. As was already noted by DeWitt, this constituted a formidable task, since a fermionic kinetic term is linear in time derivatives, and so a field-space metric that includes fermions did not seem to exist. Although DeWitt was first to note that bosons and fermions could give rise to a field-space supermanifold that includes Grassmannian coordinates, the identification of a proper metric for such a supermanifold remained elusive. In fact, this problem was so serious that it has prevented the VDW formalism from becoming as widespread in the existing literature as it deserves to be.

The present Ph.D. dissertation provides a rigorous and self-consistent solution to this problem. It explicitly shows how the correct supersymmetric metric can be derived from the action of the theory. The uniqueness of the metric is guaranteed for all ultralocal reparametrisations of the fields, with the condition that the field-space supermanifold (which is locally flat) should match the kinetic term of a free fermionic field theory. The suggested approach differs crucially from the more well-known one of Supersymmetry, as one does not need to postulate the existence of new particles or new Grassmannian dimensions.

Dr. Kieran Finn completed his Ph.D. at the Department of Physics and Astronomy at the University of Manchester in October 2020, under my supervision. His Ph.D. thesis gives a number of illustrative examples in order to showcase how the VDW formalism works not only for scalar theories, but also for theories with fermions and gravity. It is a beautiful scientific treatise that can pave the way for future developments in the field when applied to more realistic theories such as the Standard Model.

I believe that the following thesis makes a significant leap in advancing our knowledge towards a complete geometrization of our Micro-Cosmos.

Manchester, UK
March 2021

Prof. Apostolos Pilaftsis

Abstract

The ancient Greeks thought that all of creation should be describable in terms of geometry. In this thesis, we take a step towards realising this dream by applying the methods of differential geometry to modern ideas about particle physics and cosmology in the form of quantum field theory.

We shall achieve this using the formalism of field-space covariance, in which the degrees of freedom in a quantum field theory are treated as coordinates on a Riemannian manifold, known as the field-space manifold. This formalism allows us to describe such theories geometrically, in a way that is manifestly invariant under arbitrary choices such as the units, spacetime coordinates, or field variables used.

In this thesis, we extend the applicability of field-space covariance to quantum field theories with gravitational and fermionic degrees of freedom. We show how to construct the field-space manifold for such theories and how to equip it with a natural metric. Thus, we are now able to apply this formalism to all realistic theories of particle physics, including the Standard Model.

In addition, we show that the potential term in a quantum field theory can also be described geometrically through a process known as the Eisenhart lift. We show how, by introducing new degrees of freedom into the theory, a potential term can be recast as the curvature of field space.

Finally, we apply our geometric methods to the theory of inflation. We construct a manifold that describes the evolution of inflation geometrically as a geodesic. We show that the tangent bundle of this manifold, equipped with a natural metric, provides a finite measure on the initial conditions of inflation, which we can use to study fine tuning in these models.

Publications Related to This Thesis

The work in this thesis has appeared in the following publications:

[1] Kieran Finn, Sotirios Karamitsos, and Apostolos Pilaftsis. "Eisenhart lift for field theories". In: *Phys. Rev. D* 98.1 (2018), p. 016015. arXiv: 1806.02431 [physics.class-ph] (cit. on p. 147).

[2] Kieran Finn and Sotirios Karamitsos. "Finite measure for the initial conditions of inflation". In: *Phys. Rev. D* 99.6 (2019). [Erratum: Phys.Rev.D 99, 109901 (2019)], p. 063515. arXiv: 1812.07095 [gr-qc] (cit. on pp. 236, 238).

[3] Kieran Finn, Sotirios Karamitsos, and Apostolos Pilaftsis. "Frame Covariance in Quantum Gravity". In: *Phys. Rev. D* 102.4 (2020), p. 045014. arXiv: 1910.06661 [hep-th] (cit. on p. 51).

[4] Kieran Finn. "Initial Conditions of Inflation in a Bianchi I Universe". In: *Phys. Rev. D* 101.6 (2020), p. 063512. arXiv: 1912.04306 [gr-qc] (cit. on p. 233).

[5] Kieran Finn, Sotirios Karamitsos, and Apostolos Pilaftsis. "Frame Covariant Formalism for Fermionic Theories". In: *Eur. Phys. J.* C 81, (2021), p. 572. arXiv: 2006.05831 [hep-th].

[6] Kieran Finn, Sotirios Karamitsos, and Apostolos Pilaftsis. "Quantizing the Eisenhart Lift". In: *Phys. Rev. D* 103.6 (2021), p. 065004. arXiv: 2012.15288 [hep-th].

Acknowledgements

I would like to start by thanking my supervisor, Apostolos Pilaftsis. His guidance and insight throughout the last three years have been invaluable in completing the research in this thesis.

Also, Sotirios Karamitsos, who has been a great friend, as well as a collaborator. He was heavily involved in all of the work in this thesis, especially in understanding the quantum aspects of the Eisenhart lift. I would like to thank him for the countless hours spent discussing field-space metrics, probability measures, supermanifolds, time-dependent units, and the correct use of the Oxford comma.

I would also like to thank the rest of the Manchester theory office: Jack Holguin, Chris Shepherd, Matthew DeAngelis, Kiran Ostrolenk, Baptiste Cabouat, Abbie Keats, Mulham Hijazi, Pablo Candia da Silva, and Jack Helliwell for providing a fun and productive working environment with just the right balance between getting work done and going to the pub.

Throughout my Ph.D. I have been fortunate to attend many interesting conferences and workshops, which have allowed me to meet and discuss physics with amazing people from across the world. I would like to thank them for helping me develop my ideas and for pointing out issues that I would otherwise have missed. In particular, I would like to thank Matthew Kellet, Owen Goodwin, and Sam Brady for sharing their knowledge of supermanifolds; Daniel Martin for teaching me the finer points of Mathematica; and Josu Aurrokoetxea and David Sloan for helping me understand how to approach the initial conditions problem in inflation.

I would also like to thank Charlotte Owen, James Edholm, Amy Lloyd-Stubbs, and Leonora Donaldson-Wood from Lancaster University for making me feel like an honorary member of their group and for allowing me to attend their seminars and use their office when I couldn't make it into Manchester.

I would like to thank my family for always supporting me and believing in me. They are a source of inspiration and advice that I can always rely on no matter the situation.

Finally, I would like to thank my girlfriend, Anna Maria. She has had to put up with a lot throughout our relationship, from the four years spent travelling back and forth across the Atlantic, to the last six months being locked down in a house with me 24 hours a day. Yet despite all that, she continues to support me and make me feel loved every day. She may not understand much of what is written in this thesis, but it would not have been possible without her.

Contents

Chapter 1
Introduction

Everything's bigger in Texas.
—Old Texan saying

The laws of nature should not depend on the way we choose to describe them. We should not be able to influence the predictions of a theory or the results of an experiment just by using a different notation. This seemingly obvious fact has historically had far-reaching consequences. For example, imposing that the laws of physics not care about the way we choose to label space and time leads inevitably to Einstein's celebrated theory of relativity [1]. We shall refer to this idea as *reparametrisation invariance*.

In mathematics, in particular in the field of geometry, reparametrisation invariance is built into the formalism. For example, the mathematical description of a differentiable manifold is independent of the choice of coordinates we use to describe it and mathematical results can be written in a completely chart-independent way using tensor notation [2–4].

In physics, on the other hand, this principle can sometimes be less obvious. Many quantities in Quantum Field Theory (QFT) appear to depend non-trivially on the way we choose to write them down, directly violating reparametrisation invariance. We shall see some examples of this in Sect. 1.3 and in Appendix A. One of the goals of this thesis is to identify the causes of this parametrisation dependence and rectify it. We aim to achieve this goal by taking the lessons learned from differential geometry and applying them to QFT.

Geometry has always been an important tool in physics. Examples range from Kepler's early theory of the solar system in which the locations of the planets were dictated by nested Platonic solids [5] to more modern ideas of the compactifications of Calabi-Yau manifolds in string theory [6–8]. In this thesis we shall employ the tools of geometry once again and describe QFTs purely in terms of geometry.

K. Finn, *Geometric Approaches to Quantum Field Theory*, Springer Theses,
https://doi.org/10.1007/978-3-030-85269-6_1

Before we can solve them, we must first explore the problems at hand. For the rest of this introduction we will, therefore, discuss the ways in which theories of physics can pick up non-trivial dependencies on the parametrisation used to describe them.

1.1 Unit Independence

When we measure a quantity, what we are really doing is comparing it to a particular *unit*, which is an entirely arbitrary quantity that we use as a standard [9]. It can come from a physical artefact (such as the platinum-iridium cylinder at the International Bureau of Weights and Measures) or an experimental measurement (e.g. the duration of 9,192,631,770 periods of the radiation corresponding to the transition between the two hyperfine levels of the ground state of the caesium 133 atom).

Expressing the measurement of a quantity mathematically requires two ingredients: its numerical value with respect to some unit, and the definition of the unit itself [10]. For example, the length of a stick can be expressed as

$$\ell = l \times \ell_0, \tag{1.1}$$

where ℓ_0 is a unit of length (such as the length of the King's foot) and l is a dimensionless number describing the ratio of the length of the stick to the length of that unit.

Of course, we could instead use a different unit for length, such as the distance travelled by light in $1/299{,}792{,}458$ of a second. In this case, the length of the stick can be expressed as $\ell' = l' \times \ell'_0$. Reparametrisation invariance tells us that our choice of units will not affect the physical length, so $\ell' = \ell$ and, thus, we can write

$$l' = \frac{\ell_0}{\ell'_0} l. \tag{1.2}$$

This is simply a conversion between units, with $C \equiv \ell_0/\ell'_0$ acting as the conversion factor. Equation (1.2) might appear, for instance, as $1\,\text{ft} = 0.3048\,\text{m}$.

The most commonly used units (including all those already discussed) are based on artefacts or experimental measurements that do not change over space or time. For changes between such spacetime independent sets of units, reparametrisation invariance can be made manifest by use of the Buckingham-π theorem [11]. This theorem states that any physically meaningful equation involving dimensionful quantities can always be rewritten in terms of the dimensionless ratios of those quantities (labelled π_i). Since such ratios are automatically independent of the system of units, it follows that laws written in such a nondimensionalised manner will be manifestly invariant under spacetime independent changes of units.

The situation becomes more subtle, however, when we consider changes of units that depend on spacetime. In this case, reparametrisation invariance becomes less transparent. Any derivatives in the theory will now act not only on the dimensionless ratios, but also on the conversion factors and, thus, extreme care must be taken to transform all quantities appropriately.

Invariance under this more general change of units was first advocated by Weyl [12] and later by Dicke [13]. They realised that quantities measured at different points in spacetime can only be compared if they are dimensionless. Any dimensionful quantities cannot be compared unless they are brought to the same point in spacetime. It is, therefore, pure convention to say that such dimensionful quantities are equal if they are located at different spacetime points.

Consider, for example, the statement that an Olympic-sized swimming pool in the UK is the same length as one in Texas. How could we prove or disprove such a statement? We could take a tape-measure, measure the two pools and compare the results we get. But then we wouldn't actually be comparing their lengths. We would be comparing the dimensionless ratio of the length of the pool to the length of the tape measure. In the language of (1.1), we would be comparing l and not ℓ.

Alternatively, we could dismantle the Texan pool, ship it over to the UK and rebuild it next to the British one. Although we could now directly compare their lengths, both pools would be in the UK so this comparison would not help us prove the original statement. From this thought experiment it should be clear that there is no way to determine whether an Olympic–sized swimming pool located in Texas is the same length as one located in the UK and, thus, this statement must be purely a convention.

This means that we could, in theory, use a different convention. Let us claim instead that an Olympic-sized swimming pool in Texas is twice as long as one in the UK. Experimentally, we find that the dimensionless ratio of the length of a pool to the length of any other object doesn't change when you go to Texas and we must, therefore, conclude that, in fact, all objects are twice as big in the Lone Star State. The old adage is true – everything's bigger in Texas!

We must also conclude that Olympic swimmers are twice as fast in Texas, since there is no discernible difference in their lap times. Indeed, we find that any quantity with a length dimension D must be scaled by a factor 2^D inside the state. Provided we follow through all these conclusions thoroughly, however, we will find no inconsistencies and no disagreement with any experimental measurement.

We could define a new unit to correspond with our new convention. Let us call it the Texmetre and set it equal to one metre outside of Texas but half a metre inside. We would then find that a swimming pool in Texas measures 100 Texmetres, while a British one measures only 50 Texmetres. As argued above, the choice of units is pure convention and, thus, any physical prediction of a theory should not depend on whether that theory is expressed in metres or Texmetres.

It is important to note that neither the metre nor the Texmetre should be considered more fundamental, and it is purely convention to say that one unit is spacetime dependent and the other is not. If we consider the metre to be spacetime independent then we are using the standard convention and objects are considered to be the same size no matter where they are. Meanwhile, if we consider the Texmetre to be spacetime independent we have adopted our new convention and everything's bigger in Texas. The only thing that is unambiguously spacetime dependent is the conversion factor between these two systems of units $\mathcal{C}(\boldsymbol{x})$.

From the above discussion we see that we should generalise our ideas about unit invariance to include spacetime-dependent changes of units. However, in the standard formulation of QFTs, invariance under such spacetime-dependent changes of units is far from manifest. Making such a change of units corresponds to performing a Weyl transformation [14]

$$X \to \widetilde{X} = \mathcal{C}^{D(X)}(\boldsymbol{x})X \tag{1.3}$$

to all dimensionful quantities. Here, $\mathcal{C}$ is the (spacetime dependent) conversion factor, $\boldsymbol{x}$ are the coordinates of spacetime, X is a dimensionful quantity in the theory, and $D(X)$ is its dimension. Unless the theory is a specifically Weyl invariant theory, it will look very different after the transformation (1.3).

There have been several previous attempts to make invariance under local changes of units manifest. One method [15–17] involves adding a new scalar field χ with a mass dimension $D(\chi) = 1$ to the theory. All fields ϕ are then replaced with the combination $(M/\chi)^{D(\phi)}\phi$, where M is an arbitrary mass scale and $D(\phi)$ is the dimensionality of ϕ. The field χ, therefore, absorbs the Weyl transformations of ϕ and the resulting action is invariant under (1.3). This process is similar to the well-known Stueckelberg trick [18] for gauge theories.

Alternatively [19, 20], one can define a conformally covariant derivative

$$\mathcal{D}_\mu X \equiv \frac{\partial X}{\partial x^\mu} - \frac{D(X)}{2f}\frac{\partial f}{\partial x^\mu}X, \tag{1.4}$$

where f is the effective Planck mass in the theory, i.e. the coefficient of the Einstein Hilbert term in the action. Rewriting the the action and replacing partial derivatives $\partial X/\partial x^\mu$ with (1.4) ensures that all terms proportional to $d\,\mathcal{C}/dx^\mu$ will cancel out and, thus, the expression will be invariant under (1.3).

However, spacetime dependent changes of units are just a specific case of the more general notion of reparametrisation of QFTs. We wish to find a general approach that ensures manifest invariance under all such reparametrisations.

1.2 Reparametrisation Invariance in Quantum Field Theories

When constructing a QFT we must choose a set of fields ϕ^A, collectively denoted ϕ, in which to express our theory. However, there is nothing preventing us making a different choice and instead expressing our theory in terms of some other set of fields $\widetilde{\phi}^A$. Reparametrisation invariance tells us that such a choice is arbitrary and, therefore, the results of any calculations should be invariant under the transformation

$$\phi^A \to \widetilde{\phi}^A = \widetilde{\phi}^A(\phi). \tag{1.5}$$

We emphasise that the transformation (1.5) should be thought of as a reparametrisation, not as a physical transformation of the fields. Therefore, reparametrisation invariance should not be thought of as a physical symmetry of the theory, unlike, for example, the $SU(3) \times SU(2) \times U(1)$ symmetry that defines the Standard Model [21–23]. In general there will be no Noether current or gauge degrees of freedom associated with invariance under (1.5) and the invariance should not place any restrictions on the class of theories that are allowed.

Instead we should think of this transformation as a simple relabelling of the underlying degrees of freedom, akin to the choice of coordinates in describing spacetime or the choice of units discussed in the previous section. Therefore, the invariance or otherwise of a theory under (1.5) does not depend on the choice of theory, but only on the way it is written down.

This means that our mission to ensure reparametrisation invariance will not place any restrictions on the which theories are allowed and should not require the addition of new fields or symmetries. Instead we should seek a formalism that allows us to write down *any* theory in a way that makes invariance under (1.5) manifest.

We can see that the change of units (1.3) corresponds to a specific case of the general reparametrisation (1.5) provided the conversion factor $C(x)$ depends on spacetime only through the fields ϕ. This is the only sensible way to have a spacetime dependent change of units. Units can only be defined in terms of physical quantities or results, which themselves must depend on spacetime only through the quantum fields.

As discussed above, the results of any physically sensible QFT should not depend on which set of fields we use to describe it and should, therefore, be invariant under (1.5). However, in the standard formulation, this is not the case. The quantum corrections that a QFT receives can depend on the set of fields used to perform the calculations. We shall explore the reasons for this in more detail in Chap. 2, but for now let us consider an illustrative example.

1.3 Example: Complex Scalar Field

As an example to highlight these issues we will consider the theory of a single complex scalar field ϕ with a Lagrangian[1]

$$\mathcal{L} = \partial_\mu \phi \partial^\mu \phi - m^2 |\phi|^2 - \lambda |\phi|^4 . \tag{1.6}$$

We shall assume that $m^2 < 0$ so that we are in the symmetry broken vacuum with expectation value

$$\langle \phi \rangle = \frac{1}{\sqrt{2}} \rho_0 \equiv \sqrt{\frac{-m^2}{2\lambda}} . \tag{1.7}$$

[1] This analysis can be found in most textbooks on QFT. See, for example, [24, 25].

To analyse this theory let us decompose the complex scalar field ϕ into its real and imaginary parts

$$\phi = \frac{1}{\sqrt{2}} (\phi_1 + i\phi_2) . \tag{1.8}$$

We can then write (1.6) in terms of the two *real* scalar fields ϕ_1 and ϕ_2:

$$\mathcal{L} = \frac{1}{2}\partial_\mu\phi_1\partial^\mu\phi_1 + \frac{1}{2}\partial_\mu\phi_2\partial^\mu\phi_2 - \frac{1}{2}m^2(\phi_1^2 + \phi_2^2) - \frac{\lambda}{4}(\phi_1^2 + \phi_2^2)^2. \tag{1.9}$$

We can also express the the vacuum (1.7) as

$$\langle\phi_1\rangle = \rho_0, \qquad \langle\phi_2\rangle = 0. \tag{1.10}$$

We analyse quantum perturbations around this vacuum, setting

$$\phi_1 = \rho_0 + \delta\phi_1, \qquad \phi_2 = 0 + \delta\phi_2, \tag{1.11}$$

and we find the following set of Feynman rules:

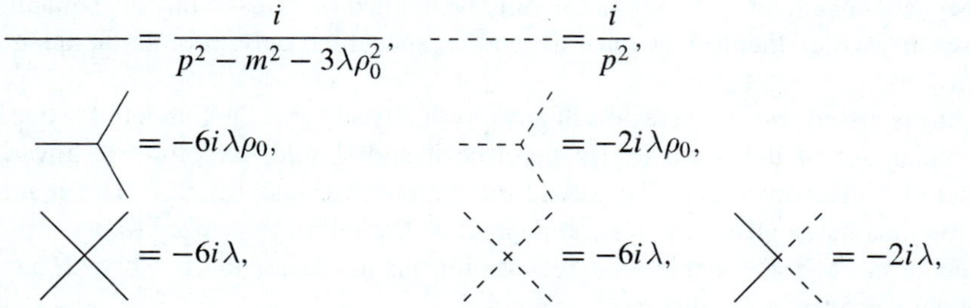

(1.12)

where a solid line corresponds to $\delta\phi_1$ and a dashed line corresponds to $\delta\phi_2$. As expected by Goldstone's theorem [26–28], we find a massless Goldstone mode corresponding to $\delta\phi_2$ and a massive Higgs mode corresponding to $\delta\phi_1$ with a mass

$$m_1^2 \equiv m^2 + 3\lambda\rho_0^2. \tag{1.13}$$

We can use these Feynman rules to calculate quantum corrections to tree level properties of the theory. For example, we can calculate the one loop correction to the self-energy of the Higgs mode, which we find to be

$i\Gamma_{\phi_1\phi_1}(p) =$ 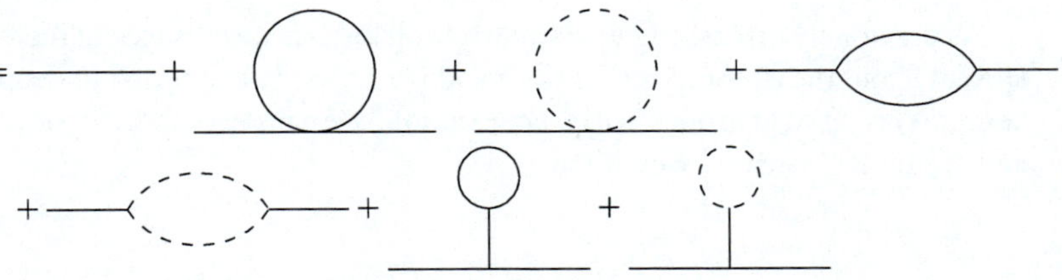

$$= (p^2 - m_1^2) + \frac{\lambda m_1^2}{4\pi^2}\ln\left(\frac{p^2}{\mu^2}\right) - \frac{\lambda m_1^2}{(4\pi)^2}\left[-4C_{\rm UV} + 4 + 9\int_0^1 dx \ln\left(\frac{x(x-1)p^2 + m_1^2}{\mu^2}\right)\right]. \tag{1.14}$$

Here, $C_{\rm UV} = \frac{2}{4-D} - \gamma_E + \ln(4\pi)$ is the ultraviolet (UV) divergence, μ is the renormalisation scale, $\gamma_E = 0.577\ldots$ is the Euler–Mascheroni constant and we have performed the calculation in $D = 4 - 2\epsilon$ dimensions using dimensional regularisation [29].

Quantum corrections can also be calculated using the effective potential [30–34]. In this formalism, quantum effects are included in the definition of the potential itself so that fully quantum results can be obtained via tree-level calculations with the effective potential. We postpone a more thorough investigation and derivation of this formalism until Chap. 2 and for now simply quote the following result. The effective potential, at one-loop order in perturbation theory, for a theory of N scalar fields ϕ^A (collectively ϕ) is given by

$$V_{\rm eff}(\phi) = \frac{1}{V_4}\frac{i}{2}\ln\det\left(\frac{\delta^2 S}{\delta\phi^A(x)\delta\phi^B(y)}\right)\bigg|_{\partial_\mu\phi=0}, \tag{1.15}$$

where V_4 is the volume of 4-space, det refers to the functional determinant and the right hand side (RHS) is to be evaluated for a static configuration.[2]

For the theory described by (1.9), we find the effective potential to be

$$\begin{aligned} V_{\rm eff}(\phi_1 = \varphi, \phi_2 = 0) = {} & \frac{1}{2}m^2\varphi^2 + \frac{1}{4}\lambda\varphi^4 \\ & + \frac{1}{64\pi^2}\left\{(m^2 + 3\lambda\varphi^2)^2\left[\ln\left(\frac{m^2 + 3\lambda\varphi^2}{\mu^2}\right) - \frac{3}{2}\right]\right. \\ & \left. + (m^2 + \lambda\varphi^2)^2\left[\ln\left(\frac{m^2 + \lambda\varphi^2}{\mu^2}\right) - \frac{3}{2}\right]\right\}, \end{aligned} \tag{1.16}$$

where we have chosen a configuration in which the field is purely real $\phi = \varphi/\sqrt{2}$ (with φ real). Because of the $U(1)$ symmetry of this theory we can make this choice without losing generality. We have removed the UV divergent part of (1.16) using the $\overline{\rm MS}$ renormalisation scheme [35, 36].

[2] If the configuration is not static then we get the effective *action*, which will be properly defined in Chap. 2.

As previously discussed, reparametrisation invariance tells us that there is nothing special about the decomposition (1.8) and we are equally entitled to decompose the complex scalar field in other ways. For example, we can decompose ϕ into its modulus and argument instead, by defining

$$\phi = \frac{1}{\sqrt{2}}\rho e^{i\frac{\sigma}{\rho_0}}. \tag{1.17}$$

We can then write (1.6) in terms of the real fields ρ and σ and we find

$$\mathcal{L} = \frac{1}{2}\partial_\mu\rho\partial^\mu\rho + \frac{1}{2}\left(\frac{\rho}{\rho_0}\right)^2 \partial_\mu\sigma\partial^\mu\sigma - \frac{1}{2}m^2\rho^2 - \frac{\lambda}{4}\rho^4. \tag{1.18}$$

The vacuum (1.7) can be expressed in these fields as

$$\langle\rho\rangle = \rho_0, \qquad \langle\sigma\rangle = 0. \tag{1.19}$$

We perform perturbation theory around this vacuum taking

$$\rho = \rho_0 + \delta\rho, \qquad \sigma = 0 + \delta\sigma. \tag{1.20}$$

Notice that, to linear order we have

$$\delta\rho = \delta\phi_1 + O(\delta\phi^2), \qquad \delta\sigma = \delta\phi_2 + O(\delta\phi^2). \tag{1.21}$$

Therefore, despite the different notation, we see that we are considering the same perturbations of the same theory around the same vacuum. Indeed, $\delta\rho$ corresponds to the massive Higgs mode with mass m_1 and $\delta\sigma$ corresponds to the massless Goldstone mode. According to reparametrisation invariance, we should, therefore, expect to obtain the same Feynman rules as we did before.

However, when we calculate the Feynman rules for (1.18) we find

 $= \frac{i}{p^2 - m_1^2}$, $= \frac{i}{p^2}$,

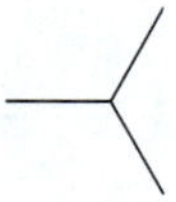 $= -6i\lambda\rho_0$, 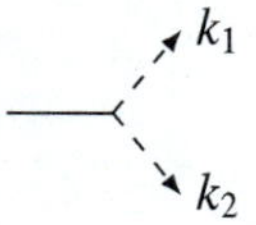 $= -\frac{2i}{\rho_0}k_1 \cdot k_2$, (1.22)

 $= -6i\lambda$, 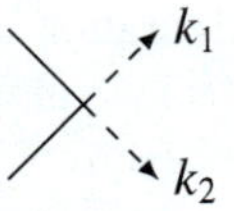 $= -\frac{2i}{\rho_0^2}k_1 \cdot k_2$,

where a solid line now refers to $\delta\rho$ and a dashed line to $\delta\sigma$. These differ from (1.12) even though the only difference in the calculations was the choice of parametrisation.

Similarly, if we again calculate the Higgs self energy, but now use ρ and σ we find

$i\Gamma_{\rho\rho}(p) =$

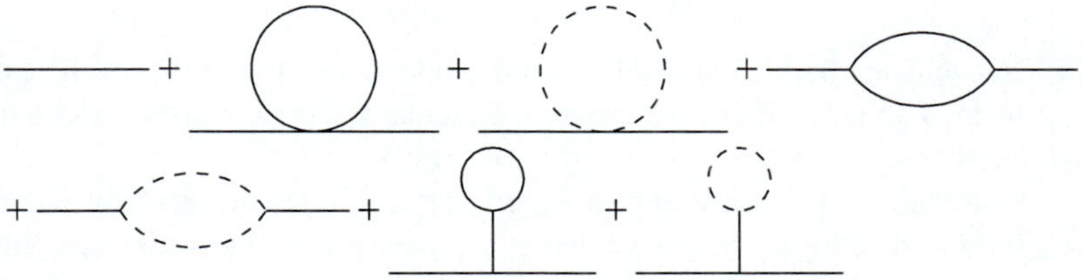

$$
\begin{aligned}
&= i(p^2 - m_1^2) + \frac{3\lambda m_1^2 + \lambda \frac{p^4}{m_1^2}}{(4\pi)^2} \ln\left(\frac{p^2}{\mu^2}\right) \\
&+ \frac{i\lambda m_1^2}{(4\pi)^2}\left[\left(3 + \frac{p^4}{m_1^4}\right) C_{\mathrm{UV}} - 6 + 2\frac{p^4}{m_1^4} - 9\int_0^1 dx \ln\left(\frac{x(x-1)p^2 + m_1^2}{\mu^2}\right)\right],
\end{aligned}
\tag{1.23}
$$

which differs from (1.14).

Finally, we can calculate the effective potential using this new parametrisation and we find

$$
\begin{aligned}
V_{\mathrm{eff}}(\rho = \varphi, \sigma = 0) &= \frac{1}{2}m^2\varphi^2 + \frac{\lambda}{4}\varphi^4 \\
&+ \frac{1}{64\pi^2}\left\{(m^2 + 3\lambda\varphi^2)^2\left[\ln\left(\frac{m^2 + 3\lambda\varphi^2}{\mu^2}\right) - \frac{3}{2}\right]\right\},
\end{aligned}
\tag{1.24}
$$

which again gives a different result to (1.16).

We find that we get different results for the Feynman rules, the one-loop Higgs self energy and the effective potential. In Appendix A.1 we show that we also get different results for the Goldstone self-energy. This is despite the fact that, in both cases, we are calculating exactly the same quantity with exactly the same perturbations around exactly the same vacuum in exactly the same theory. The only difference between the calculations (1.14) and (1.16) and the calculations (1.23) and (1.24) is the arbitrary choice of whether to express the complex scalar field in terms of its real and imaginary parts or in terms of its modulus and argument.

Before proceeding we must make an important observation. When the Higgs field is placed on shell both (1.14) and (1.16) reduce to

$$
\begin{aligned}
&i\Gamma_{\phi_1\phi_1}\left(p^2 = m_1^2\right) = i\Gamma_{\rho\rho}\left(p^2 = m_1^2\right) \\
&= -\frac{5i\lambda m_1^2}{(4\pi)^2}\ln\left(\frac{m_1^2}{\mu^2}\right) - \frac{i\lambda m_1^2}{(4\pi)^2}\left[-4C_{\mathrm{UV}} + 4 + 9\int_0^1 dx \ln\left(x^2 - x - 1\right)\right].
\end{aligned}
\tag{1.25}
$$

Similarly, the calculations in (1.16) and (1.24) both reduce to

$$V_{\text{eff}}\left(\phi=\sqrt{\frac{-m^2}{2\lambda}}\right)=-\frac{1}{4}\frac{m^4}{\lambda}+\frac{m^4}{16\pi^2}\left[\ln\left(\frac{-2m^2}{\mu^2}\right)-\frac{3}{2}\right] \tag{1.26}$$

when evaluated on the classical vacuum. Thus, whilst the results of the calculations performed above differ off shell, this difference disappears in the calculation of any on shell observable.

However, despite only appearing off shell, this parametrisation dependence is important to address. Off shell calculations have many applications in physics. One of the best examples of this is the pinch technique in gauge theories [37–40]. This technique provides a systematic way of removing gauge dependence from the off shell Green's functions of Quantum Chromodynamics (QCD). Many uses have been found for these Green's functions, including the construction of an effective charge for non-Abelian gauge theories [41], gauge-invariant definitions of the form factors in QCD [42] and the removal of gauge dependence from the Schwinger-Dyson equations [43–45]. Off shell techniques have also been shown to be useful in the analysis of quantum anomalies [46].

Furthermore, although any parametrisation dependent parts of a calculation are guaranteed to cancel when performed exactly on shell, any assumptions or approximations made during the calculation can spoil this cancellation. Much care and effort is, therefore, required to ensure on shell results remain reparametrisation invariant in the presence of such approximations. A manifestly reparametrisation invariant construction of QFT avoids these issues.

Finally, it has been shown [47–49] that when gravity is treated as a dynamical field rather than a fixed background, aspects of this parametrisation dependence can start to manifest themselves in on shell observables. This observation is known as the cosmological frame problem [50] and we shall explore it in more detail in Chap. 3.

In addition to these practical benefits, understanding this parametrisation dependence is important for the theoretical development of QFT. The cancellation of parametrisation dependent contributions to on shell observables is very delicate and any extension beyond our current understanding could easily spoil them. Resolving these issues would, therefore, not only give us a deeper knowledge of the theoretical structure of current theories of physics, but also give us greater confidence when extending these theories to describe new phenomena.

Therefore, despite only affecting off shell quantities, the differences uncovered above are unacceptable and fly in the face of reparametrisation invariance. Here we have a result from a theory that seems to depend on the way we choose to write that theory down. We take this example as motivation for the work in this thesis. Our first goal is to understand why and in what contexts such differences occur and to work out what can be done to fix it. As we shall see, this will lead us to a geometric description of QFT that has applications far beyond this simple example.

1.4 Outline

This thesis is laid out as follows.

In Chap. 2 we introduce the notion of the field space – a geometric description of QFT. We begin by reviewing several features of Riemannian geometry that will be used frequently throughout this thesis. We then review how we can construct a manifold, known as the field space, for scalar field theories and how this manifold can be used to analyse whether or not a given expression will be reparametrisation invariant. We show that many of the standard techniques used in QFT yield parametrisation dependent results, before showing how to remove this dependence through covariant Feynman diagrams and the Vilkovisky–DeWitt effective action.

In Chap. 3 we generalise the concept of field space covariance to scalar-tensor theories of gravity. We show how the interplay between the field space manifold and the spacetime manifold can lead to violations of reparametrisation invariance, which is known as the cosmological frame problem. We show how these problems can be solved through the introduction of a new model function $\ell(\phi)$ that acts as a coefficient of proportionality between the tensor field $g_{\mu\nu}$ and the metric of spacetime $\bar{g}_{\mu\nu}$. Finally, we show how this model function can be used to formulate quantum gravity in a fully reparametrisation invariant way.

In Chap. 4 we complete the field space covariance formalism by including fermionic degrees of freedom. We show how to extend the field space manifold into a supermanifold with anticommuting coordinates. We then show how to define a metric for this supermanifold from the action of a fermionic theory. We can, therefore, construct a reparametrisation invariant expression for quantum corrections to such theories.

In Chap. 5 we show how the potential term in the Lagrangian of a QFT can also be interpreted geometrically. We review a technique from classical mechanics, known as the Eisenhart lift, that allows a conservative force to be described instead in terms of the curvature of a higher-dimensional space. We show how the same technique can be applied to a QFT to rewrite the potential term in the Lagrangian as the kinetic term of a new field. This kinetic term can then be incorporated into the field space and, hence, can be interpreted geometrically.

In Chap. 6 we review the classic puzzles of Big Bang Cosmology and how the theory of inflation purports to solve them. We then review efforts to understand the initial conditions required for inflation to begin and discuss whether these initial conditions are finely tuned or not.

In Chap. 7 we show how the field space covariant formalism, together with the Eisenhart lift, can be used to study the initial conditions of inflation in a geometric way. We construct the phase space manifold for a theory of a single inflaton field in a Friedmann Robertson Walker (FRW) universe. We show that the total volume of this manifold is finite for a large class of inflationary models and, thus, show how it can be analysed to distinguish finely tuned from generic initial conditions. We also show that generalising the FRW spacetime to include anisotropy yields a negligible difference in results.

Finally, we conclude with a discussion of our findings in Chap. 8.

1.5 Conventions

Throughout this thesis we shall use units in which[3]

$$\hbar = c = k_B = M_P = 1, \tag{1.27}$$

where $\hbar$ is Planck's constant, c is the speed of light in vacuo, k_B is Boltzmann's constant,

$$M_P^2 \equiv \frac{1}{8\pi G} \tag{1.28}$$

is the reduced Planck mass and G is Newton's constant.

In addition, we shall adopt the Einstein summation notation in which repeated indices imply summation.

Finally, we shall adopt the mostly minus convention when considering the metric of spacetime and, thus, take the Minkowski metric to be

$$\eta_{\mu\nu} = \begin{pmatrix} 1 & 0 & 0 & 0 \\ 0 & -1 & 0 & 0 \\ 0 & 0 & -1 & 0 \\ 0 & 0 & 0 & -1 \end{pmatrix}. \tag{1.29}$$

References

1. Einstein A (1915) The field equations of gravitation. Sitzungsber Preuss Akad Wiss Berlin (Math Phys) 1915:844–847 (cit. on pp. 19, 39, 138, 174)
2. Ricci MMG, Levi-Civita T (1900) Méthodes de calcul différentiel absolu et leurs applications. Mathematische Annalen 54(1):125–201 (cit. on pp. 19, 41, 42)
3. Lee J (2012) Introduction to smooth manifolds. Graduate texts in mathematics. Springer New York (cit. on pp. 19, 39)
4. Petersen P (2016) Riemannian geometry. Graduate texts in mathematics. Springer International Publishing (cit. on pp. 19, 39)
5. Kepler J, Kopernikus N, Mästlin M, Schöner J (1596) Prodromus dissertationum cosmographicarum. lat. Tubingae: excudebat Georgius Gruppenbachius, p 181 (cit. on p. 19)
6. Calabi E (1954) The space of Kähler metrics. In: Proceedings of the international congress of mathematicians, pp 206–207 (cit. on p. 19)
7. Calabi E (1957) On Kähler manifolds with vanishing canonical class. Algebraic geometry and topology. A symposium in honor of S. Lefschetz. Princeton University Press, Princeton, N. J., pp 78–89 (cit. on p. 19)

[3] We temporarily re-instate $\hbar$ in Chap. 2 in order to identify different perturbation orders in QFT.

8. Yau S-T (1978) On the ricci curvature of a compact kähler manifold and the complex monge-ampére equation, I. Commun Pure Appl Math 31(3):339–411 (cit. on p. 19)
9. Bridgman PW (1922) Dimensional analysis. Yale University Press (cit. on p. 20)
10. Maxwell JC (1873) A treatise on electricity and magnetism. A treatise on electricity and magnetism v. 1. Clarendon Press (cit. on p. 20)
11. Buckingham E (1914) On physically similar systems; illustrations of the use of dimensional equations. Phys Rev 4:345–376 (cit. on p. 21)
12. Weyl H (1918) Gravitation and electricity. Sitzungsber Preuss Akad Wiss Berlin (Math Phys), p 465 (cit. on p. 21)
13. Dicke RH (1962) Mach's principle and invariance under transformation of units. Phys Rev 125:2163–2167 (cit. on pp. 21, 81)
14. Weyl H, Ehlers J, (1993) Raum, Zeit, Materie: Vorlesungen über allgemeine Relativitätstheorie. Springer (cit. on p. 23)
15. Codello A, D'Odorico G, Pagani C, Percacci R (2013) The renormalization group and Weyl-invariance. Class Quant Grav 30:115015. arXiv: 1210.3284 [hep-th] (cit. on p. 23)
16. Bars I, Steinhardt P, Turok N (2014) Local conformal symmetry in physics and cosmology. Phys Rev D89:043515. arXiv: 1307.1848 [hep-th] (cit. on p. 23)
17. Boyle L, Finn K, Turok N (2018) The big bang, CPT, and neutrino dark matter. arXiv: 1803.08930 [hep-ph] (cit. on p. 23)
18. Stueckelberg ECG (1938) Interaction energy in electrodynamics and in the field theory of nuclear forces. Helv Phys Acta 11:225–244 (cit. on p. 23)
19. Burns D, Karamitsos S, Pilaftsis A (2016) Frame-covariant formulation of inflation in scalar-curvature theories. Nucl Phys B907:785–819. arXiv: 1603.03730 [hep-ph] (cit. on pp. 23, 80, 83)
20. Karamitsos S, Pilaftsis A (2018) Frame covariant nonminimal multifield inflation. Nucl Phys B927:219–254. arXiv: 1706.07011 [hep-ph] (cit. on pp. 23, 80, 81, 83)
21. Yang CN, Mills RL (1954) Conservation of isotopic spin and isotopic gauge invariance. Phys Rev 96:191–195 (cit. on p. 24)
22. Glashow SL (1961) Partial-symmetries of weak interactions. Nucl Phys 22(4):579–588 (cit. on p. 24)
23. Weinberg S (1967) A model of leptons. Phys Rev Lett 19:1264–1266 (cit. on p. 24)
24. Peskin ME, Schroeder DV (1995) An introduction to quantum field theory. Frontiers in physics. Avalon Publishing (cit. on pp. 25, 59, 162)
25. Schwartz MD (2014) Quantum field theory and the standard model. quantum field theory and the standard model. Cambridge University Press (cit. on p. 25)
26. Nambu Y (1960) Quasiparticles and gauge invariance in the theory of superconductivity. Phys Rev 117:648–663, Taylor JC (ed) (cit. on pp. 26, 262)
27. Goldstone J (1961) Field theories with superconductor solutions. Nuovo Cim 19:154–164 (cit. on pp. 26, 262)
28. Goldstone J, Salam A, Weinberg S (1962) Broken symmetries. Phys Rev 127:965–970 (cit. on pp. 26, 262)
29. Hooft G, Veltman MJG (1972) Regularization and renormalization of Gauge fields. Nucl Phys B 44:189–213 (cit. on pp. 27, 263)
30. Sauter F (1931) Uber das Verhalten eines Elektrons im homogenen elektrischen Feld nach der relativistischen Theorie Diracs. Z Phys 69:742–764 (cit. on pp. 27, 61)
31. Heisenberg W, Euler H (1936) Consequences of Dirac's theory of positrons. Z Phys 98(11–12):714–732. arXiv: physics/0605038 [physics] (cit. on pp. 27, 61)
32. Weisskopf V (1936) The electrodynamics of the vacuum based on the quantum theory of the electron. Kong Dan Vid Sel Mat Fys Med 14N6:1–39 (cit. on pp. 27, 61)
33. Schwinger JS (1951) On gauge invariance and vacuum polarization. Phys Rev 82:664–679 (cit. on pp. 27, 61)
34. Toms DJ (2007) The Schwinger action principle and effective action. Cambridge monographs on mathematical physics. Cambridge University Press (cit. on pp. 27, 61)

35. Hooft G (1973) Dimensional regularization and the renormalization group. Nucl Phys B 61:455–468 (cit. on pp. 27, 263)
36. Weinberg S (1973) New approach to the renormalization group. Phys Rev D 8:3497–3509 (cit. on pp. 27, 263)
37. Cornwall JM, Papavassiliou J (1989) Gauge invariant three gluon vertex in QCD. Phys Rev D 40:3474 (cit. on p. 30)
38. Papavassiliou J (1990) Gauge invariant proper selfenergies and vertices in Gauge theories with broken symmetry. Phys Rev D 41:3179 (cit. on p. 30)
39. Degrassi G, Kniehl BA, Sirlin A (1993) Gauge invariant formulation of the S, T, and U parameters. Phys Rev D 48:3963–3966 (cit. on p. 30)
40. Binosi D, Papavassiliou J (2009) Pinch technique: theory and applications. Phys Rept 479:1–152. arXiv: 0909.2536 [hep-ph] (cit. on p. 30)
41. Watson NJ (1997) The Gauge independent QCD effective charge. Nucl Phys B 494:388–432. arXiv: hep-ph/9606381 (cit. on p. 30)
42. Bernabeu J, Papavassiliou J, Vidal J (2004) The Neutrino charge radius is a physical observable. Nucl Phys B 680:450–478. arXiv: hep-ph/0210055 (cit. on p. 30)
43. Binosi D, Papavassiliou J (2007) Pinch technique for Schwinger- Dyson equations. JHEP 03:041. arXiv: hep-ph/0611354 (cit. on p. 30)
44. Binosi D, Papavassiliou J (2008) Gauge-invariant truncation scheme for the Schwinger-Dyson equations of QCD. Phys Rev D 77:061702. arXiv: 0712.2707 [hep-ph] (cit. on p. 30)
45. Binosi D, Papavassiliou J (2008) New Schwinger-Dyson equations for non- Abelian gauge theories. JHEP 11:063. arXiv: 0805.3994 [hep-ph] (cit. on p. 30)
46. Grigore DR (2010) Off-shell fields and quantum anomalies. Ann U Craiova Phys 21:S117–S130 arXiv: 1011.3219 [hep-th] (cit. on p. 30)
47. Kunstatter G (1986) Vilkovisky's unique effective action: an introduction and explicit calculation. In: NATO advanced research workshop on superfield theories Vancouver, British Columbia, Canada, July 25–August 6, 1986, pp 503–517 (cit. on pp. 30, 51, 83, 90, 91, 96)
48. Kunstatter G, Leivo HP (1987) On the reparametrization dependence of the effective potential. Phys Lett B 183:75–80 (cit. on pp. 30, 51, 83, 90, 91, 96)
49. Falls K, Herrero-Valea M (2019) Frame (In)equivalence in quantum field theory and cosmology. Eur Phys J C 79(7):595. arXiv: 1812.08187 [hep-th] (cit. on pp. 30, 83, 84, 86, 97, 245)
50. Karamitsos S, Pilaftsis A (2018) On the cosmological frame problem. PoS CORFU2017:036. arXiv: 1801.07151 [hep-th] (cit. on pp. 31, 83)

Chapter 2
Field Space Covariance

God always geometrises.
—Plato

We have seen how off shell calculations in QFT can sometimes depend on the parametrisation of the fields. In this chapter we will introduce an important tool for understanding when and why this dependence arises, known as field space covariance. We will then show how this tool will allow us to reformulate these calculations in a way that is fully reparametrisation invariant.

2.1 Riemannian Geometry

Before introducing field space covariance, however, let us begin by briefly reviewing the basic properties of differential and Riemannian geometry that will be needed to develop this formalism.

Physicists commonly associate Riemannian geometry and the study differential manifolds with Einstein's theory of General Relativity [1]. However, it is important to remember that such manifolds exist as mathematical objects, independent of any connection to the spacetime. It is in this latter context that we consider geometry in this section and, indeed, throughout this thesis.

This section will provide only a brief overview of the necessary structures and mathematics required in this thesis. We refer the reader to the following textbooks for further details [2–7].

K. Finn, *Geometric Approaches to Quantum Field Theory*, Springer Theses,
https://doi.org/10.1007/978-3-030-85269-6_2

2.1.1 Manifolds

An n-dimensional manifold $\mathcal{M}$ is a topological space that can be mapped locally onto $\mathbb{R}^n$ [8]. More precisely, for each point on the manifold there exists a homeomorphism between a neighbourhood of that point and n-dimensional Euclidean space.

This homeomorphism allows us to label a point p on the manifold by n real numbers, known as the *coordinates* of p and denoted by x^μ, where $\mu \in \{1, \ldots, n\}$ (or collectively by $\boldsymbol{x}$). Such a mapping from $\mathcal{M} \leftrightarrow \mathbb{R}^n$ is known as a *chart*.

In general a single chart will not be able to cover the entire manifold and, therefore, we need a collection of several charts, known as an *atlas*. For example, we can consider the surface of the Earth to be a manifold and label the points by their latitude and longitude. Such a chart is able to uniquely label all points on the Earth except the North and South poles, which are multiply covered by the chart. Therefore, to obtain a unique label for every point on Earth we need an atlas of at least two charts.

The choice of atlas is not unique. There are an infinite number of different ways of mapping $\mathcal{M} \leftrightarrow \mathbb{R}^n$ and, hence, an infinite number of different charts. This is true even for very simple manifolds. For example, the $2D$ plane can be described using either Cartesian coordinates x and y or polar coordinates r and θ.

A simple, yet important, observation is that the choice of chart does not affect the mathematical structure of the manifold. It is simply a way to label the different points. In fact, it is possible to analyse a manifold mathematically without ever even talking about coordinates [9].

However, since coordinates are such a useful tool for describing physical phenomena, manifolds employed in physics are almost always expressed using some particular chart. It is, therefore, important to use a formalism that does not depend on this arbitrary choice of chart.

A change of chart is known as a *diffeomorphism* of the manifold and, hence, an expression that is independent of the choice of chart is said to be *diffeomorphism invariant*.

2.1.2 Tensors

Diffeomorphism invariance can be made manifest using covariant tensors on the manifold. A rank $\binom{a}{b}$ tensor is any object that transforms as [10]

$$T^{\mu_1\cdots\mu_a}_{\nu_1\cdots\nu_b} \to \widetilde{T}^{\mu_1\cdots\mu_a}_{\nu_1\cdots\nu_b} = T^{\rho_1\cdots\rho_a}_{\sigma_1\cdots\sigma_b} J^{\mu_1}_{\ \rho_1} \cdots J^{\mu_a}_{\ \rho_a} \left(J^{-1}\right)^{\sigma_1}_{\nu_1} \cdots \left(J^{-1}\right)^{\sigma_b}_{\nu_b} \tag{2.1}$$

under a diffeomorphism

$$x^\mu \to \tilde{x}^\mu = \tilde{x}^\mu(\boldsymbol{x}), \tag{2.2}$$

where

$$J^{\mu}_{\nu} = \frac{\partial \tilde{x}^{\mu}}{\partial x^{\nu}} \tag{2.3}$$

is the Jacobian matrix.

We can see from (2.1) that the product of a rank $\binom{a}{b}$ tensor with a rank $\binom{c}{d}$ tensor will give a rank $\binom{a+c}{b+d}$ tensor. We can also see that we can obtain a rank $\binom{a-m}{b-m}$ tensor by contracting m of the indices of a rank $\binom{a}{b}$ tensor. Finally, we see that a rank $\binom{0}{0}$ tensor, also known as a scalar, is invariant under the transformation (2.2). Thus, if we are going to use a manifold in physics we should ensure that any physical observables are described by scalars or fully contracted products of tensors so that the results are diffeomorphism invariant.

2.1.3 Connections and the Covariant Derivative

In order to employ tensors in physics, we need to be able to differentiate them to calculate their rate of change. However, the derivative of a tensor with respect to a coordinate does not transform as a tensor under (2.2). Explicitly we have

$$\begin{aligned} \partial_{\lambda} T^{\mu_1 \cdots \mu_a}_{\nu_1 \cdots \nu_b} \rightarrow & \partial_{\kappa} T^{\rho_1 \cdots \rho_a}_{\sigma_1 \cdots \sigma_b} \left(J^{-1}\right)^{\kappa}_{\lambda} J^{\mu_1}_{\rho_1} \cdots J^{\mu_a}_{\rho_a} \left(J^{-1}\right)^{\sigma_1}_{\nu_1} \cdots \left(J^{-1}\right)^{\sigma_b}_{\nu_b} \\ & + T^{\rho_1 \cdots \rho_a}_{\sigma_1 \cdots \sigma_b} \left(J^{-1}\right)^{\kappa}_{\lambda} \partial_{\kappa} \left(J^{\mu_1}_{\rho_1} \cdots J^{\mu_a}_{\rho_a} \left(J^{-1}\right)^{\sigma_1}_{\nu_1} \cdots \left(J^{-1}\right)^{\sigma_b}_{\nu_b}\right), \end{aligned} \tag{2.4}$$

where $\partial_{\mu} \equiv \partial/\partial x^{\mu}$. We, therefore, wish to modify the definition of the derivative so that the resulting object is still a tensor.

This leads us to define the *covariant derivative* ∇_{μ} which maps a rank $\binom{a}{b}$ tensor onto a rank $\binom{a}{b+1}$ tensor. Such a derivative can be constructed by equipping the manifold with a *connection* $\Gamma^{\mu}_{\nu\rho}$ [10–12]. Despite the index structure, a connection is not a tensor and does not transform according to (2.1). Instead, under the change of coordinates (2.2), the connection transforms as

$$\Gamma^{\mu}_{\nu\rho} \rightarrow \widetilde{\Gamma}^{\mu}_{\nu\rho} = \Gamma^{\lambda}_{\kappa\sigma} J^{\mu}_{\lambda} \left(J^{-1}\right)^{\kappa}_{\nu} \left(J^{-1}\right)^{\sigma}_{\rho} + J^{\mu}_{\lambda} \frac{\partial^2 x^{\lambda}}{\partial \tilde{x}^{\nu} \partial \tilde{x}^{\rho}}. \tag{2.5}$$

With the aid of this connection we define the covariant derivative of a tensor to be

$$\begin{aligned} \nabla_{\lambda} T^{\mu_1 \cdots \mu_a}_{\nu_1 \cdots \nu_b} = & \partial_{\lambda} T^{\mu_1 \cdots \mu_a}_{\nu_1 \cdots \nu_b} + \Gamma^{\mu_1}_{\lambda\kappa} T^{\kappa \cdots \mu_a}_{\nu_1 \cdots \nu_b} + \cdots + \Gamma^{\mu_a}_{\lambda\kappa} T^{\mu_1 \cdots \kappa}_{\nu_1 \cdots \nu_b} \\ & - \Gamma^{\kappa}_{\lambda\nu_1} T^{\mu_1 \cdots \mu_a}_{\kappa \cdots \nu_b} - \cdots - \Gamma^{\kappa}_{\lambda\nu_b} T^{\mu_1 \cdots \mu_a}_{\nu_1 \cdots \kappa}. \end{aligned} \tag{2.6}$$

By comparing (2.4) and (2.5) we see that the definition (2.6) transforms according to (2.1) under a coordinate transform (2.2) and therefore is indeed a tensor.

While the connection itself is not a tensor, we can use it to construct a tensor

$$T^{\mu}_{\nu\rho} \equiv \Gamma^{\mu}_{\nu\rho} - \Gamma^{\mu}_{\rho\nu}, \tag{2.7}$$

which is called the *torsion* [13]. The torsion can be intuitively understood as the rotation experienced by an observer living on the manifold in the absence of any external forces. In this thesis we shall only be concerned with connections for which (2.7) vanishes. We shall, therefore, not discuss the complexities of torsionful connections further. We refer the reader to [14–16] for such discussions.

2.1.4 Distances on the Manifold

Equipping a manifold with a connection is enough to define a covariant derivative (2.6) and, hence, study differential phenomena. However, it is common to also introduce a notion of distance on the manifold by defining an inner product. By introducing such an inner product we have turned the manifold into a *Riemannian* manifold [8] (or a pseudo-Riemannian manifold if the inner product is not positive definite).

An inner product can be defined in terms of an infinitesimal line element

$$ds^2 = g_{\mu\nu} dx^{\mu} dx^{\nu}. \tag{2.8}$$

Here, $g_{\mu\nu}$ is a symmetric rank $\binom{0}{2}$ tensor on the manifold, which is known as the *metric*.

A (pseudo)Riemannian manifold comes with a natural connection by virtue of the metric tensor. This is given by the Levi-Civita connection [17] (also known as a Christoffel symbol [18])

$$\Gamma^{\mu}_{\nu\rho} = \frac{1}{2} g^{\mu\sigma} \left(\partial_{\nu} g_{\sigma\rho} + \partial_{\rho} g_{\nu\sigma} - \partial_{\sigma} g_{\nu\rho} \right), \tag{2.9}$$

where $g^{\mu\nu}$ is the inverse metric satisfying $g^{\mu\rho} g_{\rho\nu} = \delta^{\mu}_{\nu}$. The connection (2.9) is the unique connection for which

$$\nabla_{\rho} g_{\mu\nu} = 0, \tag{2.10}$$

a condition known as metric compatibility. We can see that the Levi-Civita connection is symmetric in its lower two indices $\Gamma^{\mu}_{\nu\rho} = \Gamma^{\mu}_{\rho\nu}$ and is, thus, torsion-free.

Equipping the manifold with a metric also introduces the notion of volume. Using the inner product of the manifold we can calculate the volume of an infinitesimal parallelepiped with sides dx^1, $dx^2, \ldots, dx^n$ and we find it to be

$$dV = \sqrt{|g|} d^n x, \tag{2.11}$$

where $g \equiv \det(g_{\mu\nu})$ and $d^n x \equiv \prod_{i=1}^{n} dx^i$. We note that under (2.2) we have

$$\sqrt{|g|}d^n x \to \sqrt{|\tilde{g}|}d^n \tilde{x} = \left(\sqrt{|g|\,\left|\det J^{-1}\right|^2}\right)\left(|\det(J)|\,d^n x\right) = \sqrt{|g|}d^n x \tag{2.12}$$

and, thus, the volume element defined in (2.11) is diffeomorphism invariant.

2.1.5 *Curvature of a Manifold*

Manifolds that are isomorphic to Euclidean space are known as flat manifolds. For such manifolds there exists a set of coordinates, known as Cartesian coordinates, for which the metric is simply

$$g_{\mu\nu} = \delta_{\mu\nu}, \tag{2.13}$$

where $\delta_{\mu\nu}$ is the Kronecker delta.

We see that, when expressed in Cartesian coordinates, the connections on a flat manifold will vanish. Thus, the covariant derivative will be identical to the standard derivative when using these coordinates and so the commutator

$$\left[\nabla_\mu, \nabla_\nu\right] \tag{2.14}$$

will evaluate to zero. Since this commutator transforms as a tensor, it must, therefore, also give zero in any other set of coordinates on a flat manifold. Hence, a flat manifold can be identified as one for which the covariant derivative commutes.

In general, however, such Cartesian coordinates do not exist and the covariant derivative does not commute. Instead the commutator is given by

$$\begin{aligned}(\nabla_\rho\nabla_\sigma - \nabla_\sigma\nabla_\rho)T^{\mu_1\cdots\mu_a}_{\nu_1\cdots\nu_b} =& R^{\mu_1}{}_{\lambda\rho\sigma}T^{\lambda\cdots\mu_a}_{\nu_1\cdots\nu_b} + \cdots + R^{\mu_a}{}_{\lambda\rho\sigma}T^{\mu_1\cdots\lambda}_{\nu_1\cdots\nu_b} \\ & - R^{\lambda}{}_{\nu_1\rho\sigma}T^{\mu_1\cdots\mu_a}_{\lambda\cdots\nu_b} - \cdots - R^{\lambda}{}_{\nu_b\rho\sigma}T^{\mu_1\cdots\mu_a}_{\nu_1\cdots\lambda}.\end{aligned} \tag{2.15}$$

Here, we have defined the Riemann curvature tensor [19]

$$R^\rho{}_{\mu\sigma\nu} = \partial_\sigma\Gamma^\rho_{\nu\mu} - \partial_\nu\Gamma^\rho_{\sigma\mu} + \Gamma^\rho_{\sigma\lambda}\Gamma^\lambda_{\nu\mu} - \Gamma^\rho_{\nu\lambda}\Gamma^\lambda_{\sigma\mu}, \tag{2.16}$$

which can be used to quantify a manifold's departure from flatness.

The Riemann tensor is symmetric in its first and third lower indices

$$R^\rho{}_{\mu\sigma\nu} = R^\rho{}_{\nu\sigma\mu}. \tag{2.17}$$

In addition, it also satisfies the Bianchi identity [20, 21]

$$R^\rho{}_{\mu\sigma\nu} + R^\rho{}_{\sigma\nu\mu} + R^\rho{}_{\nu\mu\sigma} = 0. \tag{2.18}$$

The Riemann tensor can be contracted to construct the Ricci tensor [22]

$$R_{\mu\nu} = R^{\rho}{}_{\mu\rho\nu}, \tag{2.19}$$

which can, in turn be contracted to form the Ricci scalar

$$R = g^{\mu\nu} R_{\mu\nu}. \tag{2.20}$$

We note that, while a flat metric always has $R_{\mu\nu} = R = 0$, the converse is not always true. It is possible for the Ricci tensor or scalar to vanish, even if the manifold is curved.

2.1.6 Local Normal Coordinates and the Vielbein Formalism

Although global Cartesian coordinates only exist for flat manifolds, it is always possible to define a chart for which

$$g_{ab}(\boldsymbol{x}_*) = \delta_{ab}. \tag{2.21}$$

at a single point $\boldsymbol{x}_*$.[1] This chart is known as the local inertial frame at the point $\boldsymbol{x}_*$ and the corresponding coordinates $\tilde{x}^a(\boldsymbol{x}_*)$ (which are conventionally denoted with a lower case Latin index) are known as local normal coordinates.

The existence of local normal coordinates gives rise to a set of vectors on the manifold known as the *vielbeins* $e^a_\mu(\boldsymbol{x})$ [23, 24]. These are defined by the relation

$$\tilde{x}^a(\boldsymbol{x}_*) = e^a_\mu(\boldsymbol{x}_*)\, x^\mu. \tag{2.22}$$

The vielbeins can be schematically thought of as the "square root" of the metric. Indeed, we can infer from (2.21) and (2.22) that

$$g_{\mu\nu}(\boldsymbol{x}) = e^a_\mu(\boldsymbol{x}) e^b_\nu(\boldsymbol{x}) \delta_{ab}. \tag{2.23}$$

Because of the relation (2.23) it is possible to define a Riemannian manifold[2] by specifying the vielbeins instead of the metric.

We note that the vielbeins are, in general, not of the form

$$e^a_\mu \neq \frac{\partial \tilde{x}^a}{\partial x^\mu} \tag{2.24}$$

[1] Note, that when applied to the spacetime manifold, (2.21) is the strong form of the equivalence principle in General Relativity.

[2] It is also possible to define vielbeins for a pseudo-Riemannian manifold. In that case the Euclidean metric δ_{ab} must be replaced by the Minkowski metric η_{ab} in (2.21) and (2.23).

and, therefore, cannot be interpreted as a Jacobian. The exception to this is on a flat manifold, where the relation (2.21) holds globally, not just at the point $\boldsymbol{x}_*$.

2.1.7 Submanifolds and Induced Metrics

A *submanifold* is a restricted part of the manifold satisfying a certain set of constraints. The dimensionality of a submanifold on an n-dimensional manifold satisfying m linearly independent constraint equations is $n - m$.

Provided the constraint equations are sufficiently well-behaved, a submanifold of a Riemannian manifold is still a Riemannian manifold. The metric on the submanifold is known as the induced metric and is given by

$$\mathfrak{g}_{\tilde{\mu}\tilde{\nu}} = \frac{\partial F^{\mu}}{\partial y^{\tilde{\mu}}} \frac{\partial F^{\nu}}{\partial y^{\tilde{\nu}}} g_{\mu\nu}. \tag{2.25}$$

Here, $y^{\tilde{\mu}}$ with $\tilde{\mu} \in \{1, \ldots, n-m\}$ is a chart on the submanifold and the functions $F^{\mu}(\boldsymbol{y})$ determine how the submanifold is embedded into the parent manifold through the relation

$$x^{\mu} = F^{\mu}(\boldsymbol{y}). \tag{2.26}$$

Submanifolds are common throughout physics. For example, the surface of the Earth can be considered a two-dimensional submanifold embedded in the three-dimensional manifold describing the Universe. The constraint equation in this case would be $r = 6{,}371$ km, where r is the radial coordinate centred on the Earth's core.

2.1.8 The Geodesic Equation

A straight line is a well-defined notion in Euclidean space that corresponds to the shortest path between two points $\boldsymbol{x}_A$ and $\boldsymbol{x}_B$. We can generalise the concept of a straight line to Riemannian manifolds by defining it to be the path that minimises the total distance $\int_{\boldsymbol{x}_A}^{\boldsymbol{x}_B} ds$ between these two points. Such a path is known as a *geodesic*. In the absence of any external force, we would expect any object living on the manifold to follow one of these geodesics just as a free particle in Euclidean space always follows a straight line.

Let us derive the equation that describes such paths. We shall introduce a parameter λ and thus parametrise the curve by $\boldsymbol{x}(\lambda)$. We can normalise λ so that $\boldsymbol{x}(0) = \boldsymbol{x}_A$ and $\boldsymbol{x}(1) = \boldsymbol{x}_B$. The total length we wish to minimise can, therefore, be written

$$L[\boldsymbol{x}(\lambda)] = \int_0^1 \sqrt{g_{\mu\nu} \frac{dx^{\mu}}{d\lambda} \frac{dx^{\nu}}{d\lambda}}\, d\lambda. \tag{2.27}$$

We can treat $L[\boldsymbol{x}(\lambda)]$ as a functional of $\boldsymbol{x}(\lambda)$ and, therefore, minimize it using functional differentiation

$$\frac{\delta L}{\delta x^{\mu}(\lambda)} = \frac{1}{2\sqrt{g_{\gamma\sigma}\frac{dx^{\gamma}}{d\lambda}\frac{dx^{\sigma}}{d\lambda}}}\left[\partial_{\mu} g_{\rho\nu}\frac{dx^{\rho}}{d\lambda}\frac{dx^{\nu}}{d\lambda} - 2\partial_{\rho} g_{\mu\nu}\frac{dx^{\rho}}{d\lambda}\frac{dx^{\nu}}{d\lambda} - 2g_{\mu\nu}\frac{\partial^2 x^{\nu}}{\partial\lambda^2}\right] = 0. \tag{2.28}$$

We can rearrange (2.28) to obtain a differential equation for $\boldsymbol{x}(\lambda)$

$$\frac{\partial^2 x^{\mu}}{\partial\lambda^2} + \Gamma^{\mu}_{\nu\rho}\frac{\partial x^{\nu}}{\partial\lambda}\frac{\partial x^{\rho}}{\partial\lambda} = 0. \tag{2.29}$$

This is known as the geodesic equation.

Notice that we can choose a different parametrisation of the path $\tilde{\lambda} = a\lambda + b$ for any two constants a and b, without affecting the form of (2.29). Any choice of $\tilde{\lambda}$ satisfying this relation is known as an *affine parameter*.

2.1.9 Isometries

An *isometry* is a mapping from $\mathcal{M} \to \mathcal{M}$ that preserves the form of the metric. An isometry is, therefore, a specific type of diffeomorphism $\boldsymbol{x} \to \tilde{\boldsymbol{x}}$ that satisfies

$$\tilde{g}_{\mu\nu}(\tilde{\boldsymbol{x}}) = (J^{-1})^{\rho}_{\mu}(J^{-1})^{\sigma}_{\nu} g_{\rho\sigma}(\tilde{\boldsymbol{x}}) = g_{\mu\nu}(\boldsymbol{x}). \tag{2.30}$$

Note that in the above equation the first equality holds for any diffeomorphism due to the tensorial nature of $g_{\mu\nu}$, whereas the second equality holds only for diffeomorphisms that are also isometries.

Let us consider an infinitesimal diffeomorphism

$$x^{\mu} \to \tilde{x}^{\mu} = x^{\mu} + \xi^{\mu}(\boldsymbol{x}), \tag{2.31}$$

where[3] $\xi^{\mu} \ll 1$ is an infinitesimal vector. The Jacobian for such a transformation is

$$J^{\mu}_{\nu} = \delta^{\mu}_{\nu} + \partial_{\nu}\xi^{\mu}. \tag{2.32}$$

From (2.30) we see that the transformation (2.31) is an isometry if and only if

$$g_{\mu\nu} + \partial_{\rho} g_{\mu\nu}\xi^{\rho} - g_{\mu\rho}\partial_{\nu}\xi^{\rho} - g_{\rho\nu}\partial_{\mu}\xi^{\rho} = g_{\mu\nu}. \tag{2.33}$$

This equation can be rearranged into a more compact form

[3] Note that the coordinates x^{μ}, and hence the vector ξ^{μ} are dimensionless quantities. Thus this relation is well defined.

$$\nabla_\mu \xi_\nu + \nabla_\nu \xi_\mu = 0, \tag{2.34}$$

where we have defined $\xi_\mu = g_{\mu\nu}\xi^\nu$. Equation (2.34) is known as Killing's equation and the corresponding solutions ξ^μ are called Killing vectors [25].

2.2 The Field Space

Let us now turn our attention to QFT and try to address some of the questions raised in the Introduction. In this chapter we shall restrict ourselves to scalar field theories and, therefore, consider a theory of N scalar fields ϕ^A labelled with an index A that runs $A \in \{1, \ldots, N\}$. We shall refer to the fields collectively by ϕ. The most general Lagrangian for such a theory with up to quadratic kinetic terms is [26]

$$\mathcal{L} = \frac{1}{2} k_{AB}(\phi) g^{\mu\nu} \partial_\mu \phi^A \partial_\nu \phi^B - V(\phi), \tag{2.35}$$

where $k_{AB}(\phi)$ is the kinetic mixing matrix and $V(\phi)$ is the potential.

We have chosen to express the Lagrangian (2.35) using a particular parametrisation of the fields ϕ^A. However, as noted in the Introduction, we are perfectly entitled to choose a different set of fields $\widetilde{\phi}^A$ and, hence, write

$$\mathcal{L} = \frac{1}{2} \widetilde{k}_{AB}(\widetilde{\phi}) g^{\mu\nu} \partial_\mu \widetilde{\phi}^A \partial_\nu \widetilde{\phi}^B - \widetilde{V}(\widetilde{\phi}). \tag{2.36}$$

Here, we have the relations

$$\widetilde{k}_{AB}(\widetilde{\phi}) = (J^{-1})^C_{\ A} (J^{-1})^D_{\ B}\, k_{CD}(\phi), \tag{2.37}$$

$$\widetilde{V}(\widetilde{\phi}) = V(\phi), \tag{2.38}$$

where

$$J^A_{\ B} = \frac{\partial \widetilde{\phi}^A}{\partial \phi^B} \tag{2.39}$$

is the Jacobian of the transformation

$$\phi^A \to \widetilde{\phi}^A = \widetilde{\phi}^A(\phi). \tag{2.40}$$

The choice of whether to parametrise the theory in terms of ϕ^A or $\widetilde{\phi}^A$ is arbitrary. We, therefore, do not expect this choice to affect any calculations. However, as shown in Sect. 1.3, this is not always the case. Some off shell calculations yield different results for different parametrisations of the same theory. Let us try to understand why these differences arise by using the techniques from differential geometry that we have just discussed.

The choice of field parametrisation is strongly analogous to the choice of chart on a manifold. The fields ϕ^A are just a way of labelling the underlying degrees of freedom in the QFT, just as the coordinates are a way of labelling the underlying degrees of freedom of a manifold. In both cases the choice is arbitrary and does not affect the underlying mathematics of the theory in question, but is merely a convenient way to write the theory down. We have seen how invariance under the choice of chart can be made manifest in the case of manifolds by using tensors, so let us see if we can employ the same techniques to make reparametrisation invariance manifest in QFT.

To this end we define a manifold called the *field space* [27–32]. We take the manifold to be N dimensional and treat the fields ϕ^A as coordinates. With this identification we see that a field redefinition (2.40) is simply a diffeomorphism of the field space manifold. We can, therefore, impose reparametrisation invariance of the field theory in the same way we imposed diffeomorphism invariance in Sect. 2.1, namely by expressing our theory in terms of tensors and ensuring all indices are fully contracted.

We see already from (2.37) and (2.38) that k_{AB} is a rank $\binom{0}{2}$ tensor on the field space and V is a scalar. Similarly, $\partial_\mu \phi^A$ is a $\binom{1}{0}$ field-space tensor[4] (i.e. a field space vector) and, hence, the Lagrangian (2.35) is a field space scalar and is not dependent on the choice of field parametrisation, as expected.

In order to make the field space a Riemannian manifold we must equip it with a metric. Such a metric should satisfy the following properties [31, 32]:

1. The metric should transform as a symmetric rank $\binom{0}{2}$ tensor under (2.40).
2. The metric should be determined solely and uniquely from the Lagrangian.
3. The metric should be flat for a canonically normalised theory.

We can understand the need for these properties as follows. Property 1 is a basic requirement for the metric of any manifold as discussed in Sect. 2.1.4. Property 2 ensures that the field space is fully determined by the field theory it is trying to describe and there is no need to introduce new, ad hoc structures. Finally, property 3 means that there will be no field space effects for canonically normalised theories. This will ensure that free fields are still fully detached from the rest of the field content and can be integrated out as usual.

The Lagrangian (2.35) already has a quantity that satisfies all three of these properties – the kinetic matrix $k_{AB}(\phi)$ – and so this is what is used in the literature [31–37]. Usually $k_{AB}(\phi)$ is identified by reading it off the Lagrangian or by considering the coefficient of the highest order derivative in the Hessian matrix $H_{AB} \equiv \partial^2 \mathcal{L}/(\partial\phi^A \partial\phi^B)$. However, in [38, 39] we introduced the following equation, which allows the metric to be defined constructively:

$$G_{AB} = \frac{g_{\mu\nu}}{4} \frac{\partial^2 \mathcal{L}}{\partial(\partial_\mu \phi^A)\partial(\partial_\nu \phi^B)} - \frac{\partial^2 \mathcal{L}}{\partial\phi^A \partial(\Box\phi^B)} - \frac{\partial^2 \mathcal{L}}{\partial\phi^B \partial(\Box\phi^A)}. \tag{2.41}$$

[4] In addition to being a rank $\binom{0}{1}$ spacetime tensor (also known as a spacetime covector).

Such a constructive prescription of the field space metric is important because it ensures that the definition is unique. It also allows us to generalise to theories whose Lagrangians are not of the form (2.35), for example higher derivative theories.[5] In addition, the definition (2.41) does not depend on total derivatives of the form $\partial_\mu\big(X_A(\phi)\partial^\mu\phi^A\big)$ appearing in the Lagrangian, which do not contribute to the action. For the Lagrangian (2.35), the definition (2.41) gives

$$G_{AB} = k_{AB}. \tag{2.42}$$

With the field space metric defined we can construct the Levi-Civita connection on the field space

$$\Gamma^A_{BC} = \frac{1}{2}G^{AD}\left(\partial_B G_{DC} + \partial_C G_{BD} - \partial_D G_{BC}\right) \tag{2.43}$$

and, hence, the field-space covariant derivative

$$\begin{aligned} \nabla_C T^{A_1\cdots A_a}_{B_1\cdots B_b} = \partial_C T^{A_1\cdots A_a}_{B_1\cdots B_b} + \Gamma^{A_1}_{CD} T^{D\cdots A_a}_{B_1\cdots B_b} + \cdots + \Gamma^{A_a}_{CD} T^{A_1\cdots D}_{B_1\cdots B_b} \\ - \Gamma^D_{CB_1} T^{A_1\cdots A_a}_{D\cdots B_b} - \cdots - \Gamma^D_{CB_b} T^{A_1\cdots A_a}_{B_1\cdots D}. \end{aligned} \tag{2.44}$$

Here, we have introduced the notation $\partial_A \equiv \partial/\partial\phi^A$, which we shall use throughout this chapter.

Note that, as discussed in Sect. 2.1.3, the definition (2.43) is not the only possible choice for the connection. We could, instead, choose to equip our field space manifold with a different connection, unrelated to the metric. We could also allow for non-zero field space torsion. However, we shall find that the definition (2.43) is sufficient to ensure reparametrisation invariance. In addition, this choice of connection satisfies the metric compatibility condition

$$\nabla_A G_{BC} = 0, \tag{2.45}$$

and so will greatly simplify our calculations. We shall, therefore, not consider more general definitions of the field space connection, since they only serve to complicate the formalism.

2.2.1 *Interpretation of the Field Space*

Before moving on, let us discuss the relation between the physics of the QFT (2.35) and the geometry of the field space we have just defined.

The relation is strongest for free homogeneous fields. If we take the fields ϕ^A to be homogeneous and the potential to be $V(\phi) = 0$ then (2.35) reduces to

---

[5] In the case of a higher derivative theory, (2.41) would lead to a Finslerian metric [40] that depends on both the fields and their derivatives.

$$\mathcal{L} = \frac{1}{2} G_{AB} \dot{\phi}^A \dot{\phi}^B, \tag{2.46}$$

where a dot $\dot{}$ signifies differentiation with respect to time and we have used (2.42). We see that in this case the Lagrangian is proportional to the field-space line element

$$d\sigma^2 = G_{AB} d\phi^A d\phi^B. \tag{2.47}$$

Therefore, the equations of motion for this theory (which aim to minimise $\mathcal{L}$) will be simply the geodesic equations of the field space manifold. Indeed, applying the Euler-Lagrange equations to (2.46) yields

$$G_{AB} \ddot{\phi}^B + \frac{1}{2} \partial_C (G_{AB}) \dot{\phi}^B \dot{\phi}^C - \partial_A (G_{BC}) \dot{\phi}^B \dot{\phi}^C = 0, \tag{2.48}$$

which can be rewritten in the more familiar form

$$\ddot{\phi}^A + \Gamma^A_{BC} \dot{\phi}^B \dot{\phi}^C = 0. \tag{2.49}$$

If we now add a non-zero potential $V(\phi)$ to (2.46) we find that the equations of motion become

$$\ddot{\phi}^A + \Gamma^A_{BC} \dot{\phi}^B \dot{\phi}^C = -G^{AB} \partial_B V. \tag{2.50}$$

Therefore, we see that the potential acts as an external conservative force on the field space. In Chap. 5 we shall see that this force can also be interpreted geometrically.

Let us now generalise to the full four-dimensional QFT and so once again consider the Lagrangian (2.35). Applying the Euler-Lagrange equations we find the equations of motion are

$$\Box \phi^A + \Gamma^A_{BC} \partial_\mu \phi^B \partial^\mu \phi^C = -G^{AB} \partial_B V. \tag{2.51}$$

As shown in [26, 41, 42], if the RHS were zero, this would be the equation of motion for a three-dimensional membrane moving freely in the N-dimensional field space. The non-zero potential term on the RHS, again acts as an external force.

Another relation between the physics of the field theory and the geometry of the field space arises from its symmetries. If we consider an infinitesimal transformation of the fields

$$\phi^A \to \widetilde{\phi}^A = \phi^A + \xi^A \tag{2.52}$$

then the Lagrangian (2.35) transforms as

$$\mathcal{L} \to \widetilde{\mathcal{L}} = \mathcal{L} + \delta \mathcal{L}, \tag{2.53}$$

where

$$\delta \mathcal{L} = \left(\frac{1}{2} \partial_C (G_{AB}) \partial_\mu \phi^A \partial^\mu \phi^B - \partial_C V \right) \xi^C + G_{AC} \partial_B \xi^C \partial_\mu \phi^A \partial^\mu \phi^B. \tag{2.54}$$

A symmetry is a specific field transformation for which the model functions k_{AB} and V do not change and so $\delta\mathcal{L} = 0$. Symmetries are hugely important to QFTs because each symmetry implies the existence of a conserved current in the theory, as first shown by Noether [43]. From (2.54) we see that such a transformation must satisfy [44–47]

$$\xi^A \partial_A V = 0, \tag{2.55}$$

$$G_{AC}\partial_B \xi^C + G_{CB}\partial_A \xi^C + \xi^C \partial_C G_{AB} = 0. \tag{2.56}$$

Let us ignore (2.55) for the moment, which simply states that ξ^A must lie on a contour line of V, and focus instead on (2.56). By defining $\xi_A \equiv G_{AB}\xi^B$, this equation can be rewritten in a more familiar form:

$$\nabla_A \xi_B + \nabla_B \xi_A = 0. \tag{2.57}$$

This is simply Killing's equation (2.34) for the field space. Thus, we see that in the absence of a potential term there is a one-to-one correspondence between Noether symmetries of the field theory and Killing vectors on the field space.

Finally, we can understand the nature of the field content by looking at the signature of the field space metric. Physical particles must have a positive definite kinetic term.[6] Thus, they will contribute only positive eigenvalues to the metric. Ghost fields, on the other hand, have negative kinetic terms and will, therefore, contribute negative eigenvalues. Finally, any non-dynamical auxiliary fields will not have a kinetic term and, thus, will contribute a zero eigenvalue to the field space metric. Therefore, the signature of the field space metric will be

$$\mathrm{sign}(G_{AB}) = (p, g, a), \tag{2.58}$$

where p is the number of physical fields, g is the number of ghost fields and a is the number of auxiliary fields.

2.3 The Configuration Space

When quantising a QFT we must consider each field not as a single operator, but as an infinite set of operators, one at each point in spacetime. Therefore, to apply the techniques of field space covariance beyond the classical level we must generalise our manifold so that each coordinate is not just representing a field in general, but the value of that field at a specific point in spacetime. This leads to the definition of the configuration space manifold, which is an infinite-dimensional manifold with coordinates

$$\phi^{\widehat{A}} \equiv \phi^A(\boldsymbol{x}_A). \tag{2.59}$$

[6] Fermions are an exception to this rule as we shall discuss in Chap. 4.

Here we have used the compact notation first introduced by DeWitt [48], whereby $\widehat{A} = \{A, \boldsymbol{x}_A\}$ is a continuous index that runs over all points in spacetime as well as all scalar fields in the theory.

Throughout this thesis we shall adopt the convention that indices with hats $\widehat{\ }$ correspond to continuous configuration-space indices. Repeated configuration space indices will be taken to imply a sum over the discreet index and integration over spacetime in accordance with the Einstein-DeWitt convention. Explicitly we therefore have

$$X_{\widehat{A}} Y^{\widehat{A}} \equiv \sum_A \int d^D x_A \sqrt{-g}\, X_A(\boldsymbol{x}_A) Y^A(\boldsymbol{x}_A), \tag{2.60}$$

where $g_{\mu\nu}$, with determinant g, is the spacetime metric.

We now wish to turn the configuration space into a Riemannian manifold and so equip it with a metric. Such a metric should continue to satisfy the properties listed in Sect. 2.2 but, in addition, we require that the configuration space metric be ultra local, i.e. it should be proportional to an undifferentiated Dirac delta function [31, 32].

We can satisfy all of these conditions with the following definition of the configuration space metric:

$$\mathcal{G}_{\widehat{A}\widehat{B}} = \frac{g_{\mu\nu}}{4} \frac{\delta^2 S}{\delta(\partial_\mu \phi^{\widehat{A}}) \delta(\partial_\nu \phi^{\widehat{B}})} = G_{AB} \delta^{(4)}(\boldsymbol{x}_A - \boldsymbol{x}_B), \tag{2.61}$$

where

$$S = \int d^4 x \sqrt{-g} \mathcal{L} \tag{2.62}$$

is the action. Here, we have defined the functional derivative with respect to a partial derivative as

$$\frac{\delta F[\partial_\mu \phi^A(x)]}{\delta(\partial_\mu \phi^A(y))} \equiv \lim_{\epsilon^A_\mu \to 0} \frac{F[\partial_\mu \phi^A(x) + \epsilon^A_\mu \delta^{(D)}(x-y)] - F[\partial_\mu \phi^A(x)]}{\epsilon^A_\mu}, \tag{2.63}$$

and have used a diffeomorphism invariant definition of the Dirac delta function, normalised such that

$$\int d^D x \sqrt{-g}\, \delta^{(D)}(x) = 1. \tag{2.64}$$

With the configuration space thus equipped with a metric, it is straightforward to define the configuration-space connection

$$\Gamma^{\widehat{A}}_{\widehat{B}\widehat{C}} \equiv \frac{1}{2} \mathcal{G}^{\widehat{A}\widehat{D}} \left[\frac{\delta \mathcal{G}_{\widehat{B}\widehat{D}}}{\delta \phi^{\widehat{C}}} + \frac{\delta \mathcal{G}_{\widehat{D}\widehat{C}}}{\delta \phi^{\widehat{B}}} - \frac{\delta \mathcal{G}_{\widehat{B}\widehat{C}}}{\delta \phi^{\widehat{D}}} \right] = \Gamma^A_{BC} \delta^{(D)}(\boldsymbol{x}_A - \boldsymbol{x}_B) \delta^{(D)}(\boldsymbol{x}_A - \boldsymbol{x}_C) \tag{2.65}$$

and, hence, the configuration space covariant derivative

$$\nabla_{\widehat{C}} T^{\widehat{A}_1\cdots\widehat{A}_a}_{\widehat{B}_1\cdots\widehat{B}_b} = \frac{\delta}{\delta\phi^{\widehat{C}}} T^{\widehat{A}_1\cdots\widehat{A}_a}_{\widehat{B}_1\cdots\widehat{B}_b} + \Gamma^{\widehat{A}_1}_{\widehat{C}\widehat{D}} T^{\widehat{D}\cdots\widehat{A}_a}_{\widehat{B}_1\cdots\widehat{B}_b} + \cdots + \Gamma^{\widehat{A}_a}_{\widehat{C}\widehat{D}} T^{\widehat{A}_1\cdots\widehat{D}}_{\widehat{B}_1\cdots\widehat{B}_b} - \Gamma^{\widehat{D}}_{\widehat{C}\widehat{B}_1} T^{\widehat{A}_1\cdots\widehat{A}_a}_{\widehat{D}\cdots\widehat{B}_b} - \cdots - \Gamma^{\widehat{D}}_{\widehat{C}\widehat{B}_b} T^{\widehat{A}_1\cdots\widehat{A}_a}_{\widehat{B}_1\cdots\widehat{D}} . \tag{2.66}$$

Finally, (2.11) shows us that we can construct an invariant volume element on the configuration space [31, 32]

$$[\mathcal{D}\mathcal{V}] = \sqrt{\det(\mathcal{G})}\,[\mathcal{D}\phi] . \tag{2.67}$$

Here, $\det(\mathcal{G})$ is the functional determinant of $\mathcal{G}_{\widehat{A}\widehat{B}}$ and $[\mathcal{D}\phi] \equiv \prod_{A,x_A} d\phi^A(x_A)$. The volume element (2.67) is an invariant measure that can be used when quantizing the fields via the Feynman path integral method [49] to ensure that such a construction is done in a reparametrisation invariant way.

2.4 Parametrisation Dependence of Standard Approaches to Quantum Field Theory

Now that we have developed the technology of field space covariance, we may use it to identify where parametrisation dependence may occur in QFT calculations. As discussed above, a calculation will be independent of the parametrisation of the fields chosen if the calculation is written in terms of field or configuration space tensors and all indices are contracted. Conversely, if a calculation employs objects that are not tensors of the field or configuration space then the result of this calculation will depend on the way the theory is written down and will, thus, violate reparametrisation invariance.

In the following two subsections we analyse how quantum corrections are usually calculated in QFT to see where the parametrisation dependence identified in Chap. 1 arises.

2.4.1 Feynman Diagrams

Let us start by analysing the calculation of perturbative quantum corrections through the method of Feynman diagrams.[7] Such corrections are calculated using S-matrix elements, which can be expressed in terms of quantum correlation functions via the LSZ reduction formula [52].

A correlation function can be calculated within the path integral formulation of QFT as

---

[7] This derivation appears in most textbooks on QFT, but our approach here most closely follows [50, 51].

$$\left\langle \phi^{\widehat{A}}\phi^{\widehat{B}}\ldots\right\rangle = \frac{\int [\mathcal{D}\phi]\,(\phi^{\widehat{A}}\phi^{\widehat{B}}\ldots)e^{\frac{i}{\hbar}S[\phi_0+\phi]}}{\int [\mathcal{D}\phi]\,e^{\frac{i}{\hbar}S[\phi_0+\phi]}}. \tag{2.68}$$

Here, ϕ_0 is an arbitrary point around which to quantise – usually taken to be the classical vacuum – and we have temporarily re-instated $\hbar$ in order to keep track of perturbation order.

The easiest way to calculate such a correlation function is through the generating functional

$$Z[\boldsymbol{J}] = \int [\mathcal{D}\phi]\,\exp\left[\frac{i}{\hbar}S[\phi_0+\phi] \;+\; J_{\widehat{A}}\phi^{\widehat{A}}\right], \tag{2.69}$$

where $\boldsymbol{J}(\boldsymbol{x})$ is an external source. In terms of this generating functional, (2.68) can be rewritten

$$\left\langle \phi^{\widehat{A}}\phi^{\widehat{B}}\ldots\right\rangle = \frac{1}{Z[\boldsymbol{0}]}\left(\frac{\delta}{\delta J_{\widehat{A}}}\frac{\delta}{\delta J_{\widehat{B}}}\ldots\right)Z[\boldsymbol{J}]\bigg|_{\boldsymbol{J}=0}. \tag{2.70}$$

We now wish to calculate (2.70) perturbatively. We, therefore, Taylor expand the action to get

$$S[\boldsymbol{\phi_0}+\boldsymbol{\phi}] = \sum_n S^{(n)}_{\widehat{A}_1\ldots\widehat{A}_n}\phi^{\widehat{A}_1}\ldots\phi^{\widehat{A}_n} \tag{2.71}$$

where

$$S^{(n)}_{\widehat{A}_1\ldots\widehat{A}_n} = \frac{1}{n!}\,\frac{\delta^n S}{\delta\phi^{\widehat{A}_1}\ldots\delta\phi^{\widehat{A}_n}}\bigg|_{\phi_0}. \tag{2.72}$$

We shall ignore the constant term $S^{(0)}$ since its effects cancel in (2.70) and we will choose our quantization point ϕ_0 such that $S^{(1)}_{\widehat{A}} = 0$. Therefore, the lowest order term in (2.71) will be quadratic and so we have

$$S[\boldsymbol{\phi_0}+\boldsymbol{\phi}] = S^{(2)}_{\widehat{A}\widehat{B}}\phi^{\widehat{A}}\phi^{\widehat{B}} + \sum_{n>2} S^{(n)}_{\widehat{A}_1\ldots\widehat{A}_n}\phi^{\widehat{A}_1}\ldots\phi^{\widehat{A}_n}. \tag{2.73}$$

We can plug this expansion into (2.69) to get

$$\begin{aligned} Z[\boldsymbol{J}] &= \int [\mathcal{D}\phi]\exp\left(\frac{i}{\hbar}\sum_{n>2} S^{(n)}_{\widehat{A}_1\ldots\widehat{A}_n}\phi^{\widehat{A}_1}\ldots\phi^{\widehat{A}_n}\right)e^{\frac{i}{\hbar}S^{(2)}_{\widehat{A}\widehat{B}}\phi^{\widehat{A}}\phi^{\widehat{B}}+J_{\widehat{A}}\phi^{\widehat{A}}} \\ &= \exp\left(\frac{i}{\hbar}\sum_{n>2} S^{(n)}_{\widehat{A}_1\ldots\widehat{A}_n}\frac{\delta}{\delta J_{\widehat{A}_1}}\ldots\frac{\delta}{\delta J_{\widehat{A}_n}}\right)\int [\mathcal{D}\phi]\,e^{\frac{i}{\hbar}S^{(2)}_{\widehat{A}\widehat{B}}\phi^{\widehat{A}}\phi^{\widehat{B}}+J_{\widehat{A}}\phi^{\widehat{A}}}. \end{aligned} \tag{2.74}$$

The functional integral is now Gaussian and so can be calculated explicitly giving

$$Z[\boldsymbol{J}] = \mathcal{N}\exp\left(\frac{i}{\hbar}\sum_{n>2} S^{(n)}_{\widehat{A}_1\ldots\widehat{A}_n}\frac{\delta}{\delta J_{\widehat{A}_1}}\ldots\frac{\delta}{\delta J_{\widehat{A}_n}}\right)\exp\left(-i\hbar J_{\widehat{A}}\Delta^{\widehat{A}\widehat{B}}J_{\widehat{B}}\right), \tag{2.75}$$

where

$$\Delta^{\widehat{A}\widehat{B}} = \left(\left. \frac{\delta^2 S}{\delta\phi^{\widehat{A}}\delta\phi^{\widehat{B}}} \right|_{\phi_0} \right)^{-1} \tag{2.76}$$

is the inverse of $S^{(2)}_{\widehat{A}\widehat{B}}$, known as the propagator, and $\mathcal{N}$ is an irrelevant normalisation factor.

We can now plug this expression into (2.70) and, expanding out the two exponentials, we find

$$\begin{aligned} \left\langle \phi^{\widehat{A}}\phi^{\widehat{B}}\dots \right\rangle = \frac{\mathcal{N}}{Z[\mathbf{0}]} \left(\frac{\delta}{\delta J_{\widehat{A}}} \frac{\delta}{\delta J_{\widehat{B}}} \cdots \right) \prod_{n>2} \left[\sum_{V_n=0}^{\infty} \frac{1}{V_n!} \left(\frac{i}{\hbar} S^{(n)}_{\widehat{A}_1\dots\widehat{A}_n} \frac{\delta}{\delta J_{\widehat{A}_1}} \cdots \frac{\delta}{\delta J_{\widehat{A}_n}} \right)^{V_n} \right] \\ \times \sum_{P=0}^{\infty} \frac{1}{P!} \left. \left(-i\hbar J_{\widehat{C}} \Delta^{\widehat{C}\widehat{D}} J_{\widehat{D}} \right)^P \right|_{J=0} . \end{aligned} \tag{2.77}$$

Note that, because this expression is evaluated at $\boldsymbol{J} = 0$, the only terms that contribute are those for which

$$E + \sum_n n V_n - 2P = 0, \tag{2.78}$$

where E is the number of fields in the correlation function.

Feynman diagrams [53] are a beautiful graphical way to keep track of these non-zero terms. We can represent each propagator by a line

$$= \Delta^{\widehat{A}\widehat{B}}, \tag{2.79}$$

and each term of the expansion (2.71) by a vertex,

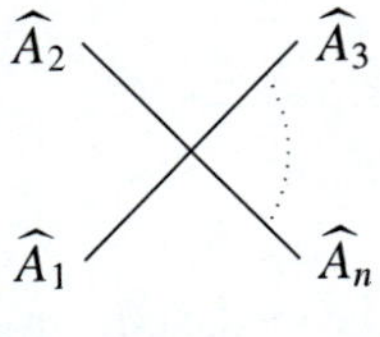

$$= n!\, S^{(n)}_{\widehat{A}_1\dots\widehat{A}_n} = \left. \frac{\delta^n S}{\delta\phi^{\widehat{A}_1}\dots\delta\phi^{\widehat{A}_n}} \right|_{\phi_0} . \tag{2.80}$$

Then each non-zero term in (2.77) can be expressed as a diagram with P propagators and V_n vertices of order n. The correlation function can be calculated by summing over all possible diagrams with the correct number of external legs. It can be shown that the effect of the prefactor $\frac{\mathcal{N}}{Z[\mathbf{0}]}$ is to remove all diagrams that are not fully connected [54].

Notice that each term is proportional to

$$\hbar^{P-\sum_n V_n}. \tag{2.81}$$

Euler's formula for connected planar graphs [55], tells us that

$$P - \sum_n V_n = N_L - 1, \tag{2.82}$$

where N_L is the number of loops in the graph. Therefore, we can see that (2.68) can be expanded perturbatively by including only diagrams up to a certain number of loops.

Let us now analyse this derivation to see whether Feynman diagrams defined in this way are reparametrisation invariant. We see that both the definition of the propagator in (2.76) and the definition of the Feynman rules in (2.80) rely on the standard functional derivative, which is not a configuration-space tensor as shown in Sect. 2.1.3. Therefore, when we contract these objects together, the resulting correlation functions are not configuration-space scalars. This leads to the parametrisation dependence we saw in Sect. 1.3.

The non-covariance of Feynman diagrams originates from the expansion (2.71), which used the fields $\phi^{\widehat{A}}$ as an expansion parameter. These fields are *coordinates* of the configuration space, not tensors, and so such an expansion is not covariant. Indeed, individual terms on the RHS of (2.71) will mix into each other under a field reparametrisation and any finite truncation of the series will not be reparametrisation invariant.

We can see further evidence of non-covariance in the term $J_{\widehat{A}}\phi^{\widehat{A}}$ in (2.69). This term involves the contraction of a configuration space tensor $J_{\widehat{A}}$ with a configuration-space coordinate $\phi^{\widehat{A}}$ and so is not reparametrisation invariant. Furthermore, the measure $[\mathcal{D}\phi]$ used throughout this derivation is not the reparametrisation invariant path integral measure (2.67) and so can lead to further parametrisation dependence.

It is, therefore, no surprise that the calculations in Sect. 1.3 gave different answers for different parametrisations of the complex scalar field. They were calculated using a parametrisation dependent method. As a reminder, this parametrisation dependence is seen only off shell and does not affect physical observables. We shall see the reason for this in the following sections.

2.4.2 *The Effective Action*

Non perturbative quantum effects can be calculated with the effective action formalism [56–60].[8] With this formalism, the quantum effects of a theory are baked into the action itself so that full quantum correlators can be calculated by simply taking the tree level processes of this action.

The quantum effective action, $\Gamma[\phi]$ can be calculated with the Legendre transform

$$\frac{i}{\hbar}\Gamma[\phi] = \ln Z[\boldsymbol{J}(\phi)] - \phi^{\widehat{A}} J_{\widehat{A}}(\phi). \tag{2.83}$$

[8] In this section we derive the full effective action $\Gamma[\phi]$. However, it is often easier to work with the effective *potential*, which is defined $V_{eff}[\phi] = -\frac{1}{V_4}\Gamma[\phi = const]$, as was done in Chap. 1. Here V_4 is the 4D volume of spacetime.

Here, $\boldsymbol{J}(\phi)$ is considered a function of ϕ satisfying the condition

$$\frac{\delta \ln(Z[J])}{\delta J_{\widehat{A}}} = \phi^{\widehat{A}}. \tag{2.84}$$

We can see from (2.83) that

$$\frac{\delta \Gamma}{\delta \phi^{\widehat{A}}} = i\hbar J_{\widehat{A}}. \tag{2.85}$$

Therefore, in the absence of external sources (i.e. $\boldsymbol{J} = 0$), the minimum of the effective action gives the full quantum vacuum.

We can invert (2.83) to obtain

$$Z[\boldsymbol{J}] = \exp\left(\frac{i}{\hbar}\Gamma[\phi(\boldsymbol{J})] + J_{\widehat{A}}\phi^{\widehat{A}}(\boldsymbol{J})\right), \tag{2.86}$$

where ϕ should now be considered a function of $\boldsymbol{J}$. Since $\Gamma[\phi]$ is evaluated at its minimum in (2.86), we can rewrite this equation as

$$Z[\boldsymbol{J}] = \lim_{\hbar \to 0} \int [\mathcal{D}\phi] \exp\left(\frac{i}{\hbar}\Gamma[\phi] + J_{\widehat{A}}\phi^{\widehat{A}}\right). \tag{2.87}$$

Notice that (2.87) is identical to (2.69), except with the action S replaced with the effective action Γ and the limit $\hbar \to 0$. This is precisely what we want. The limit $\hbar \to 0$ tells us that only tree level effects will contribute and, thus, the full quantum effects of the theory are encoded in the tree-level correlators of Γ.

We can use (2.85) to eliminate $\boldsymbol{J}$ from (2.83) and, hence, derive an implicit equation for the effective action Γ. Plugging in the definition of $Z[\boldsymbol{J}]$ from (2.69) and rearranging, we find

$$\exp\left(\frac{i}{\hbar}\Gamma[\phi]\right) = \int [D\varphi] \exp\left[\frac{i}{\hbar}S[\varphi] - \frac{i}{\hbar}(\varphi^{\widehat{A}} - \phi^{\widehat{A}})\frac{\delta\Gamma}{\delta\phi^{\widehat{A}}}\right]. \tag{2.88}$$

We can treat (2.88) as an implicit definition of Γ, since it can be calculated without any reference to external sources.

The implicit equation (2.88) can be expanded perturbatively using the background field method [61]. We take

$$\Gamma[\phi] = S[\phi] + \hbar\Gamma^{(1)}[\phi] + \cdots \tag{2.89}$$

and expand $S[\varphi]$ as in (2.71). Plugging these expansions into (2.88) we find

$$
\begin{aligned}
&\exp\left(\frac{i}{\hbar}S[\phi]+i\Gamma^{(1)}[\phi]\right)\\
&=\int[D\varphi]\exp\left[\frac{i}{\hbar}S[\phi]+\frac{i}{\hbar}(\varphi^{\widehat{A}}-\phi^{\widehat{A}})\frac{\delta S}{\delta\phi^{\widehat{A}}}+\frac{i}{2\hbar}(\varphi^{\widehat{A}}-\phi^{\widehat{A}})(\varphi^{\widehat{B}}-\phi^{\widehat{B}})\frac{\delta^2 S}{\delta\phi^{\widehat{A}}\delta\phi^{\widehat{B}}}\right.\\
&\qquad\left.-\frac{i}{\hbar}(\varphi^{\widehat{A}}-\phi^{\widehat{A}})\frac{\delta S}{\delta\phi^{\widehat{A}}}+O(\hbar^0)\right].
\end{aligned}
\tag{2.90}
$$

We see that the first term on the RHS is a constant and can, therefore, be taken out of the integral. The terms linear in $\varphi^{\widehat{A}}-\phi^{\widehat{A}}$ then cancel leaving us with a Gaussian integral, which we can calculate, and, hence, obtain

$$
\exp\left(\frac{i}{\hbar}S[\phi]\right)\exp\left(i\Gamma^{(1)}[\phi]\right)=\exp\left(\frac{i}{\hbar}S[\phi]\right)\sqrt{\det\left[\frac{\delta^2 S}{\delta\phi^{\widehat{A}}\delta\phi^{\widehat{B}}}\right]}.
\tag{2.91}
$$

The one-loop correction to the effective action is, therefore, given by

$$
\Gamma^{(1)}[\phi]=-\frac{i}{2}\det\left[\frac{\delta^2 S}{\delta\phi^{\widehat{A}}\delta\phi^{\widehat{B}}}\right].
\tag{2.92}
$$

We have a defining equation for the effective action, as well as a practical way to calculate it perturbatively, but is this effective action reparametrisation invariant? The answer, as first pointed out by Vilkovisky [31, 32], is no. Geometrically, the term $\varphi^{\widehat{A}}-\phi^{\widehat{A}}$ in (2.88) is the difference of two coordinates and not a configuration space vector. Similarly, the expression (2.92) involves the standard functional derivative, which is not covariant. Hence, we will get a different result for the effective action depending on which parametrisation of the fields we choose to work with, as discovered in Sect. 1.3.

An important observation to make is that the non-covariant part of (2.88) is proportional to $\delta\Gamma/\delta\phi^{\widehat{A}}$. It therefore vanishes when the calculation is performed on shell. This formalizes the observation we made at the end of Chap. 1; parametrisation dependence of the effective action only appears off shell and, hence, all physical observables are reparametrisation invariant.

2.5 Covariant Approaches to Quantum Field Theory

Now that we have seen how the standard approach to QFT leads to parametrisation dependent results, let us review how field space covariant techniques can be used to resolve these issues.

2.5.1 Covariant Feynman Diagrams

We start by addressing the parametrisation dependence of Feynman diagrams. As stated above, the parametrisation dependence stems from the non-covariant expansion of the action in (2.71). We, therefore, wish to replace this expansion with a covariant one.

To do so, we introduce a configuration space vector $\sigma^{\widehat{A}}[\varphi, \phi]$ first used by Vilkovisky [31, 32]. The vector $\sigma^{\widehat{A}}[\varphi, \phi]$ is defined geometrically as the tangent vector to the geodesic connecting φ and ϕ, evaluated at φ. It can be calculated by solving the differential equation

$$\sigma^{\widehat{B}}[\varphi, \phi]\,\nabla_{\widehat{B}}\,\sigma^{\widehat{A}}[\varphi, \phi] \;=\; -\sigma^{\widehat{A}}[\varphi, \phi], \tag{2.93}$$

along with the boundary conditions

$$\begin{aligned} \sigma^{\widehat{A}}[\varphi, \phi]\big|_{\varphi=\phi} &= 0, \\ -\nabla_{\widehat{B}}\sigma^{\widehat{A}}[\varphi, \phi]\big|_{\varphi=\phi} &= \delta^{A}_{B}\,\delta^{(D)}(x_A - x_B) \equiv \delta^{\widehat{A}}_{\widehat{B}}, \end{aligned} \tag{2.94}$$

where the derivative acts on the first argument in all cases. The solution, expanded to second order in $\phi^A - \varphi^A$ is [31, 32]

$$\sigma^{\widehat{A}}[\varphi, \phi] = (\phi^{\widehat{A}} - \varphi^{\widehat{A}}) + \frac{1}{2}\Gamma^{\widehat{A}}_{\widehat{B}\widehat{C}}[\varphi](\phi^{\widehat{B}} - \varphi^{\widehat{B}})(\phi^{\widehat{C}} - \varphi^{\widehat{C}}) + \cdots . \tag{2.95}$$

We note that $\sigma^{\widehat{A}}[\phi_0, \phi_0 + \phi] = \phi^{\widehat{A}} + O(\phi^2)$. This makes it a good candidate to use in place of $\phi^{\widehat{A}}$ if we wish to derive a covariant formulation of Feynman diagrams. In particular, we see that the covariant correlation functions

$$\begin{aligned} &\left\langle \sigma^{\widehat{A}}[\phi_0, \phi_0 + \phi]\sigma^{\widehat{B}}[\phi_0, \phi_0 + \phi]\ldots \right\rangle \\ &\quad = \frac{\int [\mathcal{D}\boldsymbol{\sigma}]\,(\sigma^{\widehat{A}}[\phi_0, \phi_0 + \phi]\sigma^{\widehat{B}}[\phi_0, \phi_0 + \phi]\ldots)e^{\frac{i}{\hbar}S[\phi_0+\phi]}}{\int [\mathcal{D}\boldsymbol{\sigma}]\,e^{\frac{i}{\hbar}S[\phi_0+\phi]}} \end{aligned} \tag{2.96}$$

have the same pole structure as (2.68). This means that the renormalised on shell S matrix elements

$$\prod_{I=1}^{E}\lim_{k_I^2\to m_I^2}\frac{k_I^2 - m_I^2}{Z_I^{\frac{1}{2}}}\left\langle \sigma^{\widehat{A}}\sigma^{\widehat{B}}\ldots\right\rangle = \prod_{I=1}^{E}\lim_{k_I^2\to m_I^2}\frac{k_I^2 - m_I^2}{Z_I^{\frac{1}{2}}}\left\langle \phi^{\widehat{A}}\phi^{\widehat{B}}\ldots\right\rangle \tag{2.97}$$

are identical [62]. Here, E is the number of external fields in the correlation function and k_I, m_I and Z_I are the momentum, renormalised mass and wavefunction renormalisation, respectively, of particle I.

The correlation functions will not necessarily be equal off shell. But this is to be expected since we already know (2.68) is parametrisation dependent off shell. In fact, if we want this new formalism to be reparametrisation invariant, it is *necessary* that it differs from the standard results off shell.

Armed with the definition of $\sigma^{\widehat{A}}$, we can construct a covariant expansion of the action

$$S[\phi_0 + \phi] = \sum_n \widetilde{S}^{(n)}_{\widehat{A}_1 \ldots \widehat{A}_n} \sigma^{\widehat{A}_1}[\phi_0, \phi_0 + \phi] \ldots \sigma^{\widehat{A}_n}[\phi_0, \phi_0 + \phi] , \tag{2.98}$$

where

$$\widetilde{S}^{(n)}_{\widehat{A}_1 \ldots \widehat{A}_n} = \frac{1}{n!} \nabla_{(\widehat{A}_1} \ldots \nabla_{\widehat{A}_n)} S \big|_{\phi_0} , \tag{2.99}$$

and $(\cdots)$ refers to symmetrisation over all indices. Since $\sigma^{\widehat{A}}$ and $\widetilde{S}^{(n)}_{\widehat{A}_1 \ldots \widehat{A}_n}$ are both configuration space tensors, every term on the RHS of (2.98) is individually reparametrisation invariant and, therefore, we can safely perform perturbation theory in a reparametrisation invariant way, by truncating the sum.

We can define a generating functional for the correlation functions (2.96)

$$\widetilde{Z}[J] = \int [\mathcal{D}\sigma] \, e^{\frac{i}{\hbar} S[\phi_0 + \phi] + J_{\widehat{A}} \sigma^{\widehat{A}}[\phi_0, \phi_0 + \phi]} \tag{2.100}$$

so that

$$\left\langle \sigma^{\widehat{A}}[\phi_0, \phi_0 + \phi] \sigma^{\widehat{B}}[\phi_0, \phi_0 + \phi] \ldots \right\rangle = \frac{1}{\widetilde{Z}[\mathbf{0}]} \left(\frac{\delta}{\delta J_{\widehat{A}}} \frac{\delta}{\delta J_{\widehat{B}}} \ldots \right) \widetilde{Z}[J] \bigg|_{J=0} . \tag{2.101}$$

Notice that $[\mathcal{D}\boldsymbol{\sigma}] = \sqrt{\det(\mathcal{G})} \, [\mathcal{D}\phi]$ and is, thus, equal to the measure (2.67). Also the term $J_{\widehat{A}} \sigma^{\widehat{A}}$ is now the contraction of two configuration space tensors and is, hence, a configuration space scalar. Therefore, the generating functional (2.100) is reparametrisation invariant, in contrast to the one defined in (2.69).

We can now repeat the same derivation as in Sect. 2.4.1, using the covariant generating functional (2.100) and the covariant expansion (2.98), to obtain

$$\begin{aligned} \left\langle \sigma^{\widehat{A}} \sigma^{\widehat{B}} \ldots \right\rangle = \frac{\mathcal{N}}{Z[\mathbf{0}]} \left(\frac{\delta}{\delta J_{\widehat{A}}} \frac{\delta}{\delta J_{\widehat{B}}} \ldots \right) \prod_{n>2} \left[\sum_{V_n=0}^{\infty} \frac{1}{V_n!} \left(\frac{i}{\hbar} \widetilde{S}^{(n)}_{\widehat{A}_1 \ldots \widehat{A}_n} \frac{\delta}{\delta J_{\widehat{A}_1}} \ldots \frac{\delta}{\delta J_{\widehat{A}_n}} \right)^{V_n} \right] \\ \times \sum_{P=0}^{\infty} \frac{1}{P!} \left(-i\hbar J_{\widehat{C}} \widetilde{\Delta}^{\widehat{C}\widehat{D}} J_{\widehat{D}} \right)^P \bigg|_{J=0} . \end{aligned} \tag{2.102}$$

where

$$\widetilde{\Delta}^{\widehat{A}\widehat{B}} = \left(\nabla_{\widehat{A}} \nabla_{\widehat{B}} S \big|_{\phi_0} \right)^{-1} \tag{2.103}$$

is the covariant propagator.

We see that (2.102) has the same form as (2.77) but with $S^{(n)}$ replaced with $\widetilde{S}^{(n)}$ and $\Delta^{\widehat{C}\widehat{D}}$ replaced with $\widetilde{\Delta}^{\widehat{C}\widehat{D}}$. Therefore, we can continue to calculate the correlators perturbatively using Feynman diagrams, but the Feynman rules should be given by

$$= \widetilde{\Delta}^{\widehat{A}\widehat{B}}, \tag{2.104}$$

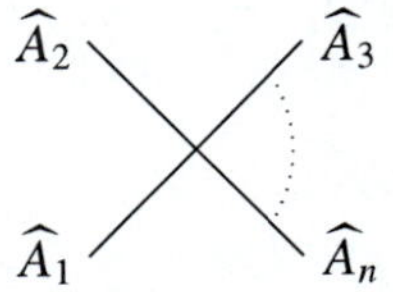

$$= \nabla_{(\widehat{A}_1} \dots \nabla_{\widehat{A}_n)} S\big|_{\phi_0} . \tag{2.105}$$

These Feynman rules are now fully covariant configuration space tensors. Therefore, the Feynman diagrams we obtain by contracting these rules are guaranteed to be reparametrisation invariant.

Note that the difference between the standard Feynman rules (2.79)–(2.80) and the covariant Feynman rules (2.104)–(2.105) is proportional to $\delta S/\delta\phi^{\widehat{A}}$ and thus vanishes on shell. This explains why the parametrisation dependence of the standard approach to QFT does not appear in physical observables.

2.5.2 The Vilkovisky–DeWitt Effective Action

Having covariantised perturbative calculations, we now wish to do the same for non-perturbative calculations by finding a covariant expression for the effective action (2.88). Such an expression was first obtained by Vilkovisky [31, 32] and was later modified by DeWitt [63] and is, hence, known as the Vilkovisky–DeWitt (VDW) effective action.

We can derive the VDW effective action by proceeding analogously to Sect. 2.4.2. However, instead of using the standard, parametrisation dependent generating function (2.69), we use the covariant version (2.100). We can, therefore, define the effective action by taking the covariant Legendre transform

$$\frac{i}{\hbar}\Gamma[\phi, \phi_*] = \ln \widetilde{Z}[\boldsymbol{J}(\phi)] - \sigma^{\widehat{A}}[\phi_*, \phi] J_{\widehat{A}}(\phi). \tag{2.106}$$

Here, as before, the current $\boldsymbol{J}$ should be considered a functional of the field ϕ satisfying

$$\frac{\delta \ln \widetilde{Z}[J]}{\delta J_{\widehat{A}}} = \sigma[\phi_*, \phi]. \tag{2.107}$$

Notice that, unlike the standard effective action, the covariant effective action depends on an arbitrary base point ϕ_* in addition to the field ϕ. This base point can also be included in the derivation of the standard effective action, but it always cancels out in the final expression (2.88) so we chose to ignore it. However, the base point does not cancel out in the derivation of the covariant effective action and so we need to include it here.

We now wish to eliminate $\boldsymbol{J}$ in order to obtain an implicit equation for the covariant effective action. We, therefore, take the derivative of (2.106), which yields

$$\frac{\delta\Gamma[\phi,\phi_*]}{\delta\phi^{\widehat{A}}}=C^{\widehat{A}}_{\widehat{B}}J^{\widehat{B}}, \tag{2.108}$$

where [63]

$$C^{\widehat{A}}_{\widehat{B}}[\varphi]\equiv-\left\langle\nabla_{\widehat{B}}\sigma^{\widehat{A}}[\varphi,\phi]\right\rangle\approx\left\langle\delta^{\widehat{A}}_{\widehat{B}}-\frac{1}{3}R^{\widehat{A}}_{\ \widehat{C}\widehat{B}\widehat{D}}[\varphi]\,\sigma^{\widehat{C}}[\varphi,\phi]\,\sigma^{\widehat{D}}[\varphi,\phi]+\ldots\right\rangle \tag{2.109}$$

and $R^{\widehat{A}}_{\ \widehat{B}\widehat{C}\widehat{D}}$ is the configuration space Riemann tensor. We note that this expression is implicit. The quantum expectation value denoted by $\langle\ \rangle$ depends on the effective action which, as we shall see, depends on the value of $C^{\widehat{A}}_{\widehat{B}}$. The only way to calculate this tensor in general is, therefore, through an iterative procedure.

Plugging (2.108) into (2.106) and using the definition (2.100) we arrive at the following implicit equation for the effective action

$$\begin{aligned}&\exp\left(\frac{i}{\hbar}\Gamma[\phi,\phi_*]\right) \\ &=\int[\mathcal{D}\varphi]\sqrt{\det(\mathcal{G})}\exp\left[\frac{i}{\hbar}S[\varphi]-\frac{i}{\hbar}\left(\Sigma^{\widehat{A}}[\phi_*,\phi]-\Sigma^{\widehat{A}}[\phi_*,\varphi]\right)\frac{\delta\Gamma}{\delta\phi^{\widehat{A}}}\right],\end{aligned} \tag{2.110}$$

where

$$\Sigma^{\widehat{A}}[\varphi,\phi]=(C^{-1}[\varphi])^{\widehat{A}}_{\widehat{B}}\,\sigma^{\widehat{B}}[\varphi,\phi]. \tag{2.111}$$

Similarly to the standard effective action, the fully quantum correlators of the field theory can be calculated from (2.110). Explicitly we have [64]

$$\left\langle\sigma^{\widehat{A}_1}\sigma^{\widehat{A}_2}\cdots\sigma^{\widehat{A}_n}\right\rangle=\nabla_{(\widehat{A}}\nabla_{\widehat{A}_2}\cdots\nabla_{\widehat{A}_n)}\Gamma[\phi,\phi_*]\big|_{\phi_*=\phi}, \tag{2.112}$$

where the covariant derivative on the RHS is with respect to the first argument ϕ and $(\cdots)$ implies symmetrisation.

We note that (2.110) still references the base point ϕ_*. This base point is arbitrary and any choice of ϕ_* yields an acceptable covariant effective action. It is, therefore, common to define

$$\Gamma_{VDW}[\phi]=\Gamma[\phi,\phi] \tag{2.113}$$

in order to simplify the algebra. This action is known as the Vilkovisky–DeWitt (VDW) effective action [31, 32, 63] (or sometimes the unique effective action) and can be obtained from the equation

$$\exp\left(\frac{i}{\hbar}\Gamma_{VDW}[\phi]\right) = \int [\mathcal{D}\varphi]\sqrt{\det(\mathcal{G})}\exp\left[\frac{i}{\hbar}S[\varphi] - \frac{i}{\hbar}\Sigma^{\widehat{A}}[\varphi,\phi]\frac{\delta\Gamma_{VDW}}{\delta\phi^{\widehat{A}}}\right]. \tag{2.114}$$

However, it is important to realise that in order to generate the covariant correlators (2.112), the derivatives must be taken *before* the identification $\phi_* = \phi$, otherwise we will introduce spurious new terms. Therefore, derivatives of the VDW effective action do not correspond to quantum correlators of the theory. This was first noted in [64, 65].

Nonetheless, we shall continue to use the VDW effective action as it presents the quantum content of a theory in a compact, reparametrisation invariant way. We note that it is straightforward to generalise to (2.110) by making the replacement

$$\Sigma^{\widehat{A}}[\varphi,\phi] \to \Sigma^{\widehat{A}}[\phi_*,\phi] - \Sigma^{\widehat{A}}[\phi_*,\varphi]. \tag{2.115}$$

Notice that when the field space is flat, the Riemann tensor vanishes and thus $C^{\widehat{A}}_{\widehat{B}} = \delta^{\widehat{A}}_{\widehat{B}}$. In this case we can replace $\Sigma^{\widehat{A}}$ with $\sigma^{\widehat{A}}$ in (2.114). This was the original definition proposed by Vilkovisky [31, 32]. For such theories, it is possible to define a parametrisation in which the fields are all canonically normalised and the field space is trivial, with Christoffel symbols $\Gamma^{\widehat{A}}_{\widehat{B}\widehat{C}} = 0$. In this parametrisation we, therefore, have $\sigma^{\widehat{A}} = \phi^{\widehat{A}}$ and hence (2.114) reduces to the standard effective action (2.88) [65].

However, for theories with non-trivial field-space curvature no such parametrisation exists and the covariant effective action must be obtained through other means such as the derivation presented above. This distinction was first noted by DeWitt [63], who introduced the full expression (2.114). He also showed that the generalisation from $\sigma^{\widehat{A}}$ to $\Sigma^{\widehat{A}}$ is required in order to satisfy

$$\left\langle \Sigma^{\widehat{A}}(\phi,\varphi)\right\rangle = 0. \tag{2.116}$$

This condition ensures that the effective action can be calculated perturbatively using only 1PI Feynman diagrams [64]. It was later shown that, in the presence of gauge symmetries, (2.116) is essential in retaining renormalisability and ensuring the resulting action is independent of the gauge fixing conditions [65, 66].

The relation (2.116) can be written diagrammatically using covariant Feynman diagrams. This gives

$(C^{-1}[\phi_0])^{\widehat{A}}_{\widehat{B}}$ 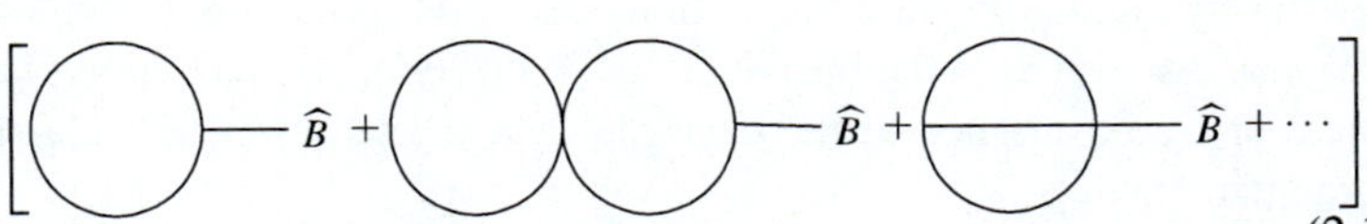 $= 0.$ (2.117)

Since this relation must hold order by order, each loop order on the left hand side (LHS) must vanish independently. In particular, this means

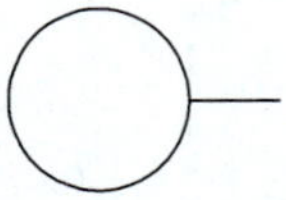
$$\widehat{A} = 0. \tag{2.118}$$

The VDW effective action can be calculated perturbatively using the background field method as before. We find

$$\Gamma_{VDW}[\phi] = S_0[\phi] + \hbar\Gamma^{(1)}[\phi] + \hbar^2\Gamma^{(2)}[\phi] + \cdots, \tag{2.119}$$

with the one and two-loop corrections being [65]

$$\Gamma^{(1)}[\phi] = -\frac{i}{2}\ln\det\nabla^{\widehat{A}}\nabla_{\widehat{B}}S, \tag{2.120}$$

$$\begin{aligned}\Gamma^{(2)}[\phi] &= \frac{1}{8}\widetilde{\Delta}^{\widehat{A}\widehat{B}}\widetilde{\Delta}^{\widehat{C}\widehat{D}}\nabla_{(\widehat{A}}\nabla_{\widehat{B}}\nabla_{\widehat{C}}\nabla_{\widehat{D})}S \\ &\quad -\frac{1}{12}\widetilde{\Delta}^{\widehat{A}\widehat{B}}\widetilde{\Delta}^{\widehat{C}\widehat{D}}\widetilde{\Delta}^{\widehat{E}\widehat{F}}\big(\nabla_{(\widehat{A}}\nabla_{\widehat{C}}\nabla_{\widehat{E})}S\big)\big(\nabla_{(\widehat{B}}\nabla_{\widehat{D}}\nabla_{\widehat{F})}S\big).\end{aligned} \tag{2.121}$$

The second order correction (2.121) can be written as a sum of two-loop covariant Feynman diagrams

$\Gamma^{(2)}[\phi] =$ 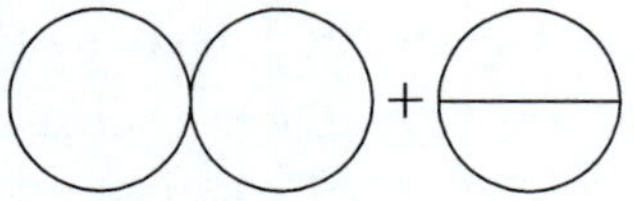 . (2.122)

Because of the condition (2.116), only the 1PI graphs contribute to $\Gamma^{(2)}[\phi]$. Other possible one-particle reducible diagrams, such as

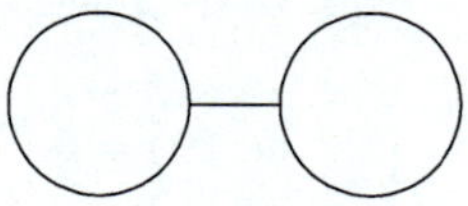
(2.123)

evaluate to zero due to (2.118) and so do not contribute.

2.6 Example: Complex Scalar Field

Let us see how the above construction solves the problems of reparametrisation invariance discovered in the Introduction. To this end, we shall apply the techniques of field space covariance to the example of a single complex scalar field ϕ with Lagrangian (1.6).

As we saw in Sect. 1.3, by expressing ϕ in terms of its real and imaginary parts we obtain the Lagrangian (1.9) for the two real fields ϕ_1 and ϕ_2. This leads to the action

$$S = \int d^4x \left[\frac{1}{2} \partial_\mu \phi_1 \partial^\mu \phi_1 + \frac{1}{2} \partial_\mu \phi_2 \partial^\mu \phi_2 - \frac{1}{2} m^2 (\phi_1^2 + \phi_2^2) - \frac{\lambda}{4} (\phi_1^2 + \phi_2^2)^2 \right]. \tag{2.124}$$

Applying (2.61) to (2.124) we see that the configuration space metric in this parametrisation is

$$\mathcal{G}_{\widehat{A}\widehat{B}}(\phi_1, \phi_2) = \begin{pmatrix} 1 & 0 \\ 0 & 1 \end{pmatrix} \delta^{(4)}(x_A - x_B). \tag{2.125}$$

This is the flat Euclidean metric, as expected for a canonically normalised theory. Therefore, the configuration space connections $\Gamma^{\widehat{A}}_{\widehat{B}\widehat{C}}$ are all zero and so the covariant Feynman rules given by (2.104)–(2.105) are identical to the regular Feynman rules (1.12). Looking at (2.95) we also see that, in this case, $\sigma^{\widehat{A}}[\varphi, \phi] = \phi^{\widehat{A}} - \varphi^{\widehat{A}}$ and so the VDW effective action is identical to the standard effective action and, thus, the VDW effective potential is given by (1.16).

Now let us consider the alternative parametrisation in terms of the modulus and argument of ϕ as we did in (1.17). In this parametrisation the action is given by

$$S = \int d^4x \left[\frac{1}{2} \partial_\mu \rho \partial^\mu \rho + \frac{1}{2} \left(\frac{\rho}{\rho_0} \right)^2 \partial_\mu \sigma \partial^\mu \sigma - \frac{1}{2} m^2 \rho^2 - \frac{\lambda}{4} \rho^4 \right]. \tag{2.126}$$

Applying (2.61) to (2.126) we see that the configuration space metric in this parametrisation is

$$\widetilde{\mathcal{G}}_{\widehat{A}\widehat{B}}(\rho, \sigma) = \begin{pmatrix} 1 & 0 \\ 0 & \left(\frac{\rho}{\rho_0} \right)^2 \end{pmatrix} \delta^{(4)}(x_A - x_B). \tag{2.127}$$

Notice that

$$\widetilde{\mathcal{G}}_{\widehat{A}\widehat{B}} = (J^{-1})^{\widehat{C}}_{\widehat{A}} (J^{-1})^{\widehat{D}}_{\widehat{B}} \mathcal{G}_{\widehat{C}\widehat{D}}, \tag{2.128}$$

where

$$(J^{-1})^{\widehat{A}}_{\widehat{B}} = \left(\frac{\partial \{\rho, \sigma\}}{\partial \{\phi_1, \phi_2\}} \right)^{-1} = \begin{pmatrix} \cos\left(\frac{\sigma}{\rho_0}\right) & \sin\left(\frac{\sigma}{\rho_0}\right) \\ \frac{\rho}{\rho_0} \sin\left(\frac{\sigma}{\rho_0}\right) & -\frac{\rho}{\rho_0} \cos\left(\frac{\sigma}{\rho_0}\right) \end{pmatrix} \delta^{(4)}(x_A - x_B) \tag{2.129}$$

is the inverse Jacobian of the transformation $\{\phi_1, \phi_2\} \to \{\rho, \sigma\}$. As expected, the configuration space metric transforms as a rank $\binom{0}{2}$ tensor.

The configuration space, written in this parametrisation, is non-trivial and has the following non-zero connections:

$$\begin{aligned} \Gamma^{\rho(z)}_{\sigma(x)\sigma(y)} &= -\frac{\rho(z)}{\rho_0^2} \delta^{(4)}(z - x) \delta^{(4)}(z - y), \\ \Gamma^{\sigma(z)}_{\rho(x)\sigma(y)} &= \frac{1}{\rho(z)} \delta^{(4)}(z - x) \delta^{(4)}(z - y). \end{aligned} \tag{2.130}$$

These non-trivial field space effects must be taken into account when calculating quantum corrections in a covariant way.

For example, we must now use (2.104)–(2.105) to calculate the covariant Feynman rules for this theory. For the action (2.126) we find these to be

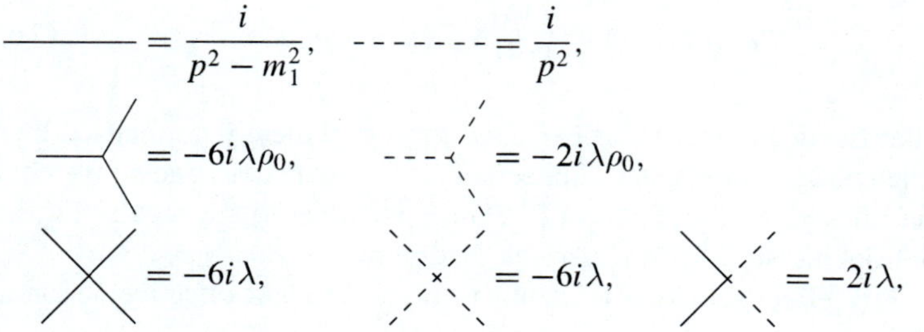

$$= \frac{i}{p^2 - m_1^2}, \quad = \frac{i}{p^2}, \quad = -6i\lambda\rho_0, \quad = -2i\lambda\rho_0, \quad = -6i\lambda, \quad = -6i\lambda, \quad = -2i\lambda, \tag{2.131}$$

where a solid line corresponds to $\delta\rho$ and a dashed line corresponds to $\delta\sigma$. These are now identical to (1.12) as required.

Similarly, we can now use the VDW effective action formalism to calculate the effective potential at one loop from (2.120). We find that, taking into account the non-trivial field space, the covariant inverse propagator evaluated for a static real configuration is given by

$$\nabla_{\hat{A}}\nabla_{\hat{B}}S\big|_{\rho=const,\sigma=0} = \begin{pmatrix} -\partial^2 - m^2 - 3\lambda\rho^2 & 0 \\ 0 & -\rho^2\partial^2 - m^2\rho^2 - \lambda\rho^4 \end{pmatrix} \delta(x_A - x_B). \tag{2.132}$$

We can, therefore, calculate the effective potential and find

$$\begin{aligned} V_{\text{eff}}(\rho = \varphi, \sigma = 0) =& \frac{1}{2}m^2\varphi^2 + \frac{1}{4}\lambda\varphi^4 \\ &+ \frac{1}{64\pi^2}\left\{(m^2 + 3\lambda\varphi^2)^2\left[\ln\left(\frac{m^2 + 3\lambda\varphi^2}{\mu^2}\right) - \frac{3}{2}\right]\right. \\ &\left. +(m^2 + \lambda\varphi^2)^2\left[\ln\left(\frac{m^2 + \lambda\varphi^2}{\mu^2}\right) - \frac{3}{2}\right]\right\}, \end{aligned} \tag{2.133}$$

in agreement with (1.16).

We see that the parametrisation dependence identified in the Introduction has been successfully removed using the methods of field space covariance. In Appendix we analyse the parametrisation dependence of this theory in more detail and perform covariant quantum calculations for other example theories.

References

1. Einstein A (1915) The field equations of gravitation. In: Sitzungsber. Preuss. Akad. Wiss. Berlin (Math Phys), pp 844–847 (cit. on pp. 19, 39, 138, 174)
2. Lee J (2012) Introduction to smooth manifolds. Graduate texts in mathematics. Springer, New York (cit. on pp. 19, 39)
3. Petersen P (2016) Riemannian geometry. Graduate texts in mathematics. Springer International Publishing (cit. on pp. 19, 39)
4. Hartle JB (2003) Gravity: an introduction to Einstein's general relativity. Addison- Wesley (cit. on p. 39)
5. Wald RM (2010) General relativity. University of Chicago Press (cit. on p. 39)
6. Misner CW, Thorne KS, Wheeler JA, Kaiser DI (2017) Gravitation. Princeton University Press (cit. on p. 39)
7. Carroll SM (2019) Spacetime and geometry. Cambridge University Press (cit. on p. 39)
8. Riemann B (1851) Grundlagen für eine allgemeine Theorie der Functionen einer veränderlichen complexen Grösse". PhD thesis. University of Göttingen (cit. on pp. 40, 43)
9. Cartan É (1899) Sur certaines expressions différentielles et le problème de Pfaff. fr. In: Annales scientifiques de l'École Normale Supérieure 3e série 16:239–332 (cit. on p. 40)
10. Ricci MMG, Levi-Civita T (1900) Méthodes de calcul différentiel absolu et leurs applications. Mathematische Annalen 54(1):125–201 (cit. on pp. 19, 41, 42)
11. Cartan E (1926) Les groupes d'holonomie des espaces généralisés. Acta Math 48(1–2):1–42 (cit. on p. 42)
12. Cartan Ë (1924) Sur les variétés à connexion projective. fr. Bulletin de la Société Mathématique de France 52:205–241 (cit. on p. 42)
13. Cartan É (1923) Sur les variétés à connexion affine et la théorie de la relativité généralisée (première partie)". fr. Annales scientifiques de l'École Normale Supérieure 3e sÉrie, 40:325–412 (cit. on p. 42)
14. Hehl FW, Von Der Heyde P, Kerlick GD, Nester JM (1976) General relativity with spin and torsion: foundations and prospects. Rev Mod Phys 48:393–416 (cit. on p. 42)
15. Strominger A (1986) Superstrings with torsion. Nucl Phys B 274:253 (cit. on p. 42)
16. Shapiro IL (2002) Physical aspects of the space-time torsion. Phys Rept 357:113. arXiv: hep--th/0103093 (cit. on p. 42)
17. Levi-Civita MT (1916) Nozione di parallelismo in una varietà qualunque e conseguente specificazione geometrica della curvatura riemanniana. Rendiconti del Circolo Matematico di Palermo (1884-1940) 42(1):173–204 (cit. on p. 43)
18. Christoffel EB (1869) Ueber die Transformation der homogenen Differentialausdrücke zweiten Grades. J für die reine und angewandte Mathematik 1869(70):46–70 (cit. on p. 43)
19. Riemann B (1867) Ueber die Hypothesen, welche der Geometrie zu Grunde liegen. Abhandlungen der Königlichen Gesellschaft der Wissenschaften in Göttingen. in der Dieterichschen Buchhandlun (cit. on p. 44)
20. Voss A (1880) Zur Theorie der Transformation quadratischer Differentialausdrücke und der Krümmung höherer Mannigfaltigkeiten. Mathematische Annalen 16(2):129–179 (cit. on p. 44)
21. Bianchi L (1902) Sui simboli a quattro indici e sulla curvatura di Riemann: nota. Tip. della R. Accademia dei Lincei (cit. on p. 44)
22. Curbastro GR (1904) Direzioni e invarianti principali in una varieta qualunque: nota. Officine grafiche Ferrari (cit. on p. 45)
23. Schwinger JS (1963) Quantized gravitational field. Phys Rev 130:1253–1258 (cit. on p. 45)
24. Yepez J (2011) Einstein's vierbein field theory of curved space. arXiv: 1106.2037 [gr-qc] (cit. on p. 45)
25. Killing W (1889) Erweiterung des Begriffes der Invarianten von Transformationsgruppen. Mathematische Annalen 35(3):423–432 (cit. on p. 48)
26. Gell-Mann M, Lévy M (1960) The axial vector current in beta decay. Il Nuovo Cimento (1955–1965) 16(4):705–726 (cit. on pp. 49, 53)

27. Honerkamp J (1972) Chiral multiloops. Nucl Phys B 36:130–140 (cit. on p. 50)
28. Honerkamp J (1972) The Question of invariant renormalizability of the massless Yang-Mills theory in a manifest covariant approach. Nucl Phys B 48:269–287 (cit. on p. 50)
29. Ecker G, Honerkamp J (1972) Covariant perturbation theory and chiral superpropagators. Phys Lett B 42:253–256 (cit. on p. 50)
30. Honerkamp J, Krause F, Scheunert M (1974) On the equivalence of standard and covariant perturbation series in non-polynomial pion lagrangian field theory. Nucl Phys B 69:618–636 (cit. on p. 50)
31. Vilkovisky GA (1984) The unique effective action in quantum field theory. Nucl Phys B234:125–137 (cit. on pp. 50, 51, 55, 56, 63, 64, 67-69, 79, 90, 91, 94, 96, 124, 243)
32. Vilkovisky GA (1984) The gospel according to DeWitt, pp 169–209 (cit. on pp. 50, 51, 55, 56, 63, 64, 67-69, 79, 90, 91, 94, 96, 124, 243)
33. Kunstatter G (1986) Vilkovisky's unique effective action: an introduction and explicit calculation. In: NATO advanced research workshop on superfield theories Vancouver, British Columbia, Canada, July 25–August 6, pp 503–517 (cit. on pp. 30, 51, 83, 90, 91, 96)
34. Kunstatter G, Leivo HP (1987) On the reparametrization dependence of the effective potential. Phys Lett B 183:75–80 (cit. on pp. 30, 51, 83, 90, 91, 96)
35. Sasaki M, Tanaka T (1998) Superhorizon scale dynamics of multiscalar inflation. Prog Theor Phys 99:763–782. arXiv: gr-qc/9801017 [gr-qc] (cit. on p. 51)
36. Groot Nibbelink S, van Tent BJW (2000) Density perturbations arising from multiple field slow roll inflation. arXiv: hep- ph/0011325 [hep-ph] (cit. on pp. 51, 173)
37. Gong J-O, Tanaka T (2011) A covariant approach to general field space metric in multi-field inflation. JCAP 1103:015. [Erratum: JCAP 1202, E01 (2012)]. arXiv: 1101.4809 [astro-ph.CO] (cit. on pp. 51, 173)
38. Kieran Finn, Sotirios Karamitsos, and Apostolos Pilaftsis. "Frame Covariance in Quantum Gravity". In: Phys. Rev. D 102.4 (2020), p. 045014. arXiv: gr-qc/1910.06661 [hep-th] (cit. on p. 51)
39. Kieran Finn, Sotirios Karamitsos, and Apostolos Pilaftsis. "Frame Covariant Formalism for Fermionic Theories". In: Eur. Phys. J. C 81, (2021), p. 572. arXiv: 2006.05831 [hep-th]
40. Finsler P (1918) Über Kurven und Flächen in allgemeinen Räumen. German. Göttingen, Zürich: O. Füssli, 120 S. 80 (cit. on p. 51)
41. Brink L, Di Vecchia P, Howe P (1976) A locally supersymmetric and reparametrization invariant action for the spinning string. Phys Lett B 65(5):471–474 (cit. on p. 53)
42. Polyakov AM (1981) Quantum geometry of bosonic strings. Phys Lett B 103(3):207–210 (cit. on p. 53)
43. Noether E (1918) Invariante Variationsprobleme. ger. Nachrichten von der Gesellschaft der Wissenschaften zu Göttingen, Mathematisch-Physikalische Klasse 1918:235–257 (cit. on p. 54)
44. de Ritis R, Marmo G, Platania G et al (1990) New approach to find exact solutions for cosmological models with a scalar field. Phys Rev D42:1091–1097 (cit. on p. 54)
45. Capozziello S, de Ritis R (1993) Relation between the potential and nonminimal coupling in inflationary cosmology. Phys Lett A177:1–7 (cit. on p. 54)
46. Capozziello S, De Ritis R, Rubano C, Scudellaro P (1996) Noether symmetries in cosmology. Riv Nuovo Cim 19N4:1–114 (cit. on p. 54)
47. Kaewkhao N, Kanesom T, Channuie P (2018) Noether symmetry approach in the cosmological alpha-attractors. Nucl Phys B931:216–225. arXiv: 1711.10080 [gr-qc] (cit. on p. 54)
48. DeWitt BS (1967) Quantum theory of gravity. 2. The manifestly covariant theory. Phys Rev 162:1195–1239 [298 (1967)] (cit. on pp. 55, 79)
49. Feynman RP (1948) Space-time approach to nonrelativistic quantum mechanics. Rev Mod Phys 20:367–387 (cit. on p. 56)
50. Srednicki M (2007) Quantum field theory. Cambridge University Press (cit. on p. 57)
51. Ryder LH (1996) Quantum field theory. Cambridge University Press (cit. on p. 57)
52. Lehmann H, Symanzik K, Zimmermann W (1955) On the formulation of quantized field theories. Nuovo Cim 1:205–225 (cit. on p. 57)

53. Feynman RP (1949) The theory of positrons. Phys Rev 76:749–759 (cit. on p. 59)
54. Peskin ME, Schroeder DV (1995) An introduction to quantum field theory. Frontiers in physics. Avalon Publishing (cit. on pp. 25, 59, 162)
55. Euler L (1758) Elementa doctrinae solidorum. Novi Commentarii academiae scientiarum Petropolitanae 4:109–140 (cit. on p. 60)
56. Sauter F (1931) Uber das Verhalten eines Elektrons im homogenen elektrischen Feld nach der relativistischen Theorie Diracs. Z Phys 69:742–764 (cit. on pp. 27, 61)
57. Heisenberg W, Euler H (1936) Consequences of Dirac's theory of positrons. Z Phys 98(11–12):714–732. arXiv: physics/0605038 [physics] (cit. on pp. 27, 61)
58. Weisskopf V (1936) The electrodynamics of the vacuum based on the quantum theory of the electron. Kong Dan Vid Sel Mat Fys Med 14N6:1–39 (cit. on pp. 27, 61)
59. Schwinger JS (1951) On gauge invariance and vacuum polarization. Phys Rev 82:664–679 (cit. on pp. 27, 61)
60. Toms DJ (2007) The schwinger action principle and effective action. Cambridge monographs on mathematical physics. Cambridge University Press (cit. on pp. 27, 61)
61. Abbott LF (1982) Introduction to the background field method. Acta Phys Polon B13:33 (cit. on p. 62)
62. Lee BW (1975) Gauge theories. Conf Proc C 7507281:79–139 (cit. on p. 64)
63. DeWitt BS (1985) The effective action. In: Architecture of fundamental interactions at short distances: proceedings, Les Houches 44th summer school of theoretical physics: Les Houches, France, July 1-August 8, pt2. 1987, pp 1023–1058 (cit. on pp. 67-69, 105, 124, 243)
64. Burgess CP, Kunstatter G (1987) On the physical interpretation of the Vilkovisky-de Witt effective action. Mod Phys Lett A2:875 [Erratum: Mod Phys Lett A2:1003 (1987)] (cit. on pp. 68, 69)
65. Ellicott P, Toms DJ (1989) On the new effective action in quantum field theory. Nucl Phys B312:700–714 (cit. on pp. 69, 70)
66. Rebhan A (1987) The Vilkovisky-de Witt effective action and its application to Yang-Mills theories. Nucl Phys B288:832–857 (cit. on pp. 69, 105)

Chapter 3
Frame Covariance in Quantum Gravity

"Time and space and gravitation have no separate existence from matter."
—Albert Einstein

In the previous chapter we introduced the notion of field space covariance for scalar field theories. In this chapter we want to extend the applicability of such techniques to include gravity. Covariant formulations of gravity have received a lot of attention [8–15]. However, a consensus has not yet been reached. As we shall see, when gravity is treated as a dynamical field, the parametrisation dependence of (2.88) can manifest itself in physical on shell calculations and not just off shell quantities.

There has been much debate as to whether this dependence is physical or just an artefact of an incomplete formalism. Based on the arguments in Chap. 1 it seems clear that no physical quantity should depend on the way it is written, and so any parametrisation dependence must be the consequence of an incomplete formalism. It is, therefore, imperative to address this incompleteness and, thus, construct a reparametrisation invariant effective action for theories with gravity.

We shall concentrate on scalar-tensor theories of gravity [16–22]. Such a theory consists of a set of N scalar fields ϕ^A (collectively denoted ϕ) and a spin-2 graviton field $g_{\mu\nu}$, and has an action of the form

$$
\begin{aligned}
S &\equiv \int d^4x \sqrt{-g}\, \mathcal{L} \ , \\
\mathcal{L} &= -\frac{f(\phi)}{2} R + \frac{1}{2} g^{\mu\nu} k_{AB}(\phi) \partial_\mu \phi^A \partial_\nu \phi^B - V(\phi),
\end{aligned}
\tag{3.1}
$$

where $f(\phi)$ is the effective Planck mass and $k_{AB}(\phi)$ and $V(\phi)$ are the scalar field-space metric and potential, respectively, as in Chap. 2. We shall refer to these three functions collectively as *model functions* [2, 3, 23].

K. Finn, *Geometric Approaches to Quantum Field Theory*, Springer Theses,
https://doi.org/10.1007/978-3-030-85269-6_3

In this chapter we shall be treating $g_{\mu\nu}$ as a dynamical field and shall quantise it accordingly. As is well known, quantising gravity renders (3.1) nonrenormalisable [24, 25]. Finding a UV complete, renormalisable theory of quantum gravity remains one of the most important unsolved problems in physics today. However, UV complete theories of quantum gravity are not the focus of this thesis and so we shall make no attempt to solve this problem. Nonetheless, we believe that the observations made in this chapter will be important to take into account when constructing a UV complete theory of quantum gravity.

3.1 The Cosmological Frame Problem

Instead we shall focus on the problem of reparametrisation invariance in scalar tensor theories. Within the context of such theories, reparametrisations take the form

$$\begin{aligned} g_{\mu\nu} &\to \tilde{g}_{\mu\nu} = \tilde{g}_{\mu\nu}(g_{\rho\sigma}, \phi), \\ \phi^A &\to \tilde{\phi}^A = \tilde{\phi}^A(g_{\rho\sigma}, \phi). \end{aligned} \tag{3.2}$$

Equation (3.2) represents the most general reparametrisations possible without mixing fields and their derivatives.[1] However, if we do not want to introduce any new spacetime tensors, we must restrict ourselves to the following reparametrisations

$$g_{\mu\nu} \to \tilde{g}_{\mu\nu} = \Omega^2(\phi) g_{\mu\nu}, \tag{3.3}$$

$$\phi^A \to \tilde{\phi}^A = \tilde{\phi}^A(\phi), \tag{3.4}$$

which are known as a conformal transformation and a scalar field redefinition, respectively. Together they constitute a frame transformation.

Under such a frame transformation, the model functions in (3.1) transform as [3]

$$f \to \tilde{f} = \Omega^{-2} f\,, \tag{3.5}$$

$$V \to \tilde{V} = \Omega^{-4} V\,, \tag{3.6}$$

$$\begin{aligned} k_{AB} \to \tilde{k}_{AB} = K^C_{\ A} K^D_{\ B} \big[& k_{CD} - 6 f \Omega^{-2} \partial_C \Omega \partial_D \Omega \\ & + 3\Omega^{-1} \partial_C f \partial_D \Omega + 3\Omega^{-1} \partial_D f \partial_C \Omega \big]\,, \end{aligned} \tag{3.7}$$

where $\partial_A \equiv \partial/\partial\phi^A$ as before and $K^A_{\ B} \equiv \partial\tilde{\phi}^A/\partial\phi^B$ is the Jacobian of the scalar field reparametrisation.

We notice from (3.5) that it is always possible to transform to a frame in which $f = 1$ by performing a conformal transformation with $\Omega = \sqrt{f}$. Such a frame is known as the *Einstein frame* because the gravitational part of the action is simply the Einstein-

[1] We shall not consider reparametrisations that depend on derivatives in this thesis, but shall comment on such transformations briefly in Chap. 8.

Hilbert action. Thus, $g_{\mu\nu}$ satisfies Einstein's equations in this frame. This is in contrast to the original frame $f \neq 1$, known as the *Jordan frame*, in which there is a non-trivial interaction between the scalar fields and the Ricci scalar and so Einstein's equations must be modified.

Scalar-tensor theories are usually specified using the Jordan frame, and the model functions are often simpler in that frame. However, one must often transform to the Einstein frame in order to make calculations. Mathematically we see that these two formulations should be equivalent as they are related by a field reparametrisation. However, there is a long-standing debate as to whether these two mathematically equivalent formulations actually correspond to the same theory of physics.

The debate was first sparked for the Brans-Dicke theory of gravity [1, 18], which is a specific case of (3.1) with a single scalar field ϕ and with Lagrangian

$$\mathcal{L} = -\frac{1}{2}\phi R + \frac{1}{2}\frac{\omega}{\phi}\partial_\mu\phi\partial^\mu\phi \tag{3.8}$$

in the Jordan frame, where ω is a constant. We see that this corresponds to the choice of model functions $f(\phi) = \phi$, $k_{AB} = \omega/\phi$ and $V(\phi) = 0$.

As noted by Faraoni and Gunzig [26], the Lagrangian (3.8) expressed in the Jordan frame is unbounded from below and is, therefore, classically unstable. Meanwhile, the same theory expressed in the Einstein frame gives a Lagrangian

$$\mathcal{L} = -\frac{1}{2}\widetilde{R} + \frac{1}{2}\partial_\mu\widetilde{\phi}\partial^\mu\widetilde{\phi}, \tag{3.9}$$

which is perfectly well behaved. Here, the Einstein and Jordan frame fields are related by

$$\widetilde{g}_{\mu\nu} = \phi g_{\mu\nu}, \qquad \widetilde{\phi} = \int \frac{\sqrt{\omega+3}}{\phi}d\phi. \tag{3.10}$$

They, therefore, concluded that only the Einstein-frame formulation was physically meaningful. A similar conclusion was also reached by Cho [27] by considering gravitational waves in the two formulations.

Meanwhile, a parallel debate exists in cosmology. The effects of frame choice on cosmology can be studied by restricting the spacetime metric $g_{\mu\nu}$ in (3.1) to the homogeneous isotropic Friedmann Robertson Walker (FRW) metric [28–34][2]

$$ds^2 = dt^2 - a(t)^2(dx^2 + dy^2 + dz^2). \tag{3.11}$$

It was found [35–39] that in this case the conformal transformation between the Jordan and Einstein frames requires a non-trivial redefinition of time to bring the metric back into the form (3.11). This time redefinition means that many of the standard cosmological quantities do not agree in the two formulations.

[2] We shall discuss the dynamics of the FRW metric in more detail in Chap. 6.

Further frame-dependencies can also crop up when matter fields are added to the theory. These matter fields can only have their standard coupling to gravity in one frame, normally taken to be the Jordan frame, and calculations performed in any other frame will take on non-standard forms. For example, the equivalence principle, which states that the gravitational and inertial masses of an object are equal, is valid only in the frame with standard couplings to matter [40, 41]. Performing the calculation in a frame with non-standard couplings to matter can also affect how the Universe is reheated after inflation [42, 43].

However, it was later shown that the two formulations are, in fact, equivalent at the classical level. Several authors [2, 3, 7, 44, 45] showed that (3.1) could be rewritten in terms of objects that are invariant under conformal transformations, thereby enforcing classical reparametrisation invariance. Furthermore, it was shown [46, 47] that all cosmological parameters that are frame dependent are also unobservable, and anything observable is frame invariant when all effects are taken into consideration.

Things become less clear-cut, however, when quantum corrections are considered. One-loop corrections to the effective action for scalar-tensor theories show an explicit difference between calculations performed in the Einstein frame and the Jordan frame [48–51]. We have already seen the origin for these differences in Chap. 2, which arise due to the parametrisation dependence of the standard effective action. It was shown [52, 53] that using the VDW effective action instead makes the calculations agree.

However, this is not the end of the story. The parametrisation-dependence discussed in Chap. 2 only appears when the calculations are performed off shell. Meanwhile, the parametrisation dependence in scalar tensor gravity is sometimes present even for on shell calculations. One of the first works to point this out was by Kunstatter and Leivo [5][3] who calculated quantum corrections in Kaluza–Klein theory [57, 58]. The authors expressed the five-dimensional Kaluza–Klein metric as

$$g_{AB} = \phi^s \begin{pmatrix} g_{\mu\nu} & \phi A_\mu \\ \phi A_\nu & \phi \end{pmatrix}, \tag{3.12}$$

where $g_{\mu\nu}$ is the 4D metric, A_μ is a 4-vector and ϕ is a scalar. They found that the quantum corrections to this theory were different if they were calculated using the 5D metric g_{AB} or the 4D quantities $g_{\mu\nu}$, A_μ and ϕ, even when the calculation was performed on shell.

This discovery was formalised for all theories of the form (3.1) by Falls and Herrero–Valea [6]. They showed that, even though the VDW effective action is inherently reparametrisation invariant once the configuration space has been defined, the definition of the configuration-space metric is frame dependent. Indeed, we see that the usual way of defining the configuration space metric, (2.61) involves the metric $g_{\mu\nu}$ explicitly, with no mention of whether this is the Jordan frame or Einstein frame metric.

---

[3] See also [4, 14, 54–56] for follow-up work.

The authors calculated one-loop quantum effects in scalar tensor theories and showed that there is a non-trivial difference between the effective action as calculated in the Jordan frame and the Einstein frame that persists even when the calculation is performed on shell. This results in a frame-dependent contribution to the effective action that they named the *frame discriminant*. They claimed that the only way to calculate quantum effects for scalar tensor theories was to choose a "preferred frame" and quantise using the configuration space metric (2.61) defined in that frame. They, therefore, claimed that the Einstein frame and Jordan frame formulations of (3.1) are different theories with different physical predictions at the quantum level.

3.2 The Invariant Spacetime Metric

The discussion in Chapters 1 and 2 suggest a different interpretation of these results. The Einstein frame and the Jordan frame are related by a field redefinition and so reparametrisation invariance tells us that they are just different expressions of the same theory. If we are getting different results in different frames it must be because our approach is incorrect, not because one frame is physically "preferred" over another.

Falls and Herrero–Valea [6] have already identified the cause of the problem, which is the factor of $g_{\mu\nu}$ appearing in (2.61). This factor is needed to ensure that the configuration space metric is a scalar under spacetime diffeomorphisms. The issues, therefore, stem from the dual nature of $g_{\mu\nu}$. We are trying to treat $g_{\mu\nu}$ as both a coordinate on our field space manifold and as the metric of the spacetime manifold.

This dual nature means that the spacetime line element

$$ds^2 = g_{\mu\nu} dx^\mu dx^\nu \tag{3.13}$$

is frame dependent. Indeed, we see that it picks up a factor of Ω^{-2} under a conformal transformation (3.3).

The frame dependence of the spacetime line element prevents us from having manifest invariance under both field reparametrisations and spacetime diffeomorphisms. Any attempt to use (3.13) to impose diffeomorphism invariance will inevitably lead to frame dependence. We should, therefore, redefine the metric of spacetime such that the spacetime line element is a scalar under diffeomorphisms of the configuration space, just as the configuration space line element is a scalar under diffeomorphisms of spacetime.

Let us, therefore, redefine the spacetime metric in the following, reparametrisation invariant, way:

$$\bar{g}_{\mu\nu} = \frac{1}{\ell^2(\phi)} g_{\mu\nu}, \tag{3.14}$$

where $\ell(\phi)$ is a newly introduced model function.

We shall refer to ℓ as the *effective Planck length*. It has dimensions of length and, thus, transforms as

$$\ell \to \widetilde{\ell} = \Omega \ell \tag{3.15}$$

under (3.3). In addition, it does not transform under (3.4) and we, therefore, see that $\bar{g}_{\mu\nu}$ is a field space scalar as required.

In particular cases ℓ could be related to the other model functions in our theory: f, k_{AB} and V. For example, previous works [59–63] have defined a similar reparametrisation invariant spacetime metric, but have assumed a particular form of ℓ, such as $\ell = 1/\sqrt{f}$. In general though, there is no reason to assume any relation between ℓ and the other model functions. Thus, ℓ is an independent model function that must be specified in addition to f, k_{AB} and V when defining a scalar tensor theory in this formalism.

Let us clarify the difference between $g_{\mu\nu}$ and $\bar{g}_{\mu\nu}$. The tensor $g_{\mu\nu}$ is a spin-2 quantum field and should be treated on the same footing as the other quantum fields in our theory ϕ^A. It is therefore $g_{\mu\nu}$ that acts as a coordinate in our field space, $g_{\mu\nu}$ that transforms under a field redefinition and $g_{\mu\nu}$ that obeys the equations of motion and appears in Feynman diagrams. This means that it is $g_{\mu\nu}$ that satisfies Einstein's equations in the Einstein frame.

On the other hand, $\bar{g}_{\mu\nu}$ is the metric of spacetime. It is invariant under field redefinitions and is not a coordinate of the field space. It is the metric $\bar{g}_{\mu\nu}$ that defines the spacetime line element

$$d\bar{s}^2 = \bar{g}_{\mu\nu} dx^\mu dx^\nu \tag{3.16}$$

and it is therefore $\bar{g}_{\mu\nu}$ that must be used in the construction of spacetime diffeomorphism invariant objects.

Note that the line element defined by (3.16) is dimensionless. This is an essential property for any quantity to be reparametrisation invariant as discussed in Sect. 1.1.

There always exists a frame in which $\ell(\phi) = 1$ in Planck units. In that frame, which we refer to as the metric frame, $\bar{g}_{\mu\nu} = g_{\mu\nu}$. But $\bar{g}_{\mu\nu}$ is frame invariant whereas $g_{\mu\nu}$ is not and, thus, they will not be equal in any other frame.

We note that the metric frame is equivalent to the "preferred frame" of Falls and Herrero-Valea [6]. Thus, specifying $\ell(\phi)$ is mathematically equivalent to choosing a preferred frame in which to quantise. However, in our formalism the metric frame is just one frame of many and is no more "preferred" than any other.

In addition, by defining the spacetime line element (3.16) in a reparametrisation invariant way, we can now fully separate the two spaces and, hence, construct a formalism that maintains manifest invariance under both spacetime diffeomorphisms and frame transformations. Let us investigate some of the consequences of this construction.

The first thing to notice is that we can now construct a reparametrisation invariant spacetime volume element

$$d\bar{V} = \sqrt{-\bar{g}}\, d^{(D)}x, \tag{3.17}$$

where $\bar{g}$ is the determinant of $\bar{g}_{\mu\nu}$. This directly affects how we take any spacetime integral, including when we contract Einstein DeWitt indices. Indeed, we see that we should now define

$$X_{\bar{A}}\, Y^{\bar{A}} = \sum_A \int d^D x_A \sqrt{-\bar{g}}\, X_A(x_A)\, Y^A(x_A). \tag{3.18}$$

We shall denote configuration space indices with a bar $\bar{}$ instead of a hat when we wish to use this convention.

The volume element (3.17) also leads us to define a frame- and diffeomorphism-invariant Dirac delta function

$$\bar{\delta}^{(D)}(x) \equiv \ell^D \delta^{(D)}(x) \tag{3.19}$$

such that

$$\int d^D x \sqrt{-\bar{g}}\, \bar{\delta}^{(D)}(x) = 1. \tag{3.20}$$

This definition, in turn, leads us to define functional differentiation as

$$\frac{\bar{\delta} F[\Phi(x)]}{\bar{\delta}\Phi(y)} \equiv \lim_{\epsilon \to 0} \frac{F[\Phi(x) + \epsilon \bar{\delta}^{(D)}(x-y)] - F[\Phi(x)]}{\epsilon}, \tag{3.21}$$

where $F[\Phi(x)]$ is a functional and $\Phi(x)$ is a field. Functional derivatives defined in this way will inherit their transformation properties from the functional F and field Φ only. This is in contrast to standard functional derivatives, which pick up extra transformation properties from $g_{\mu\nu}$.

We also see that, when taking the functional determinant of a configuration space matrix we need to use the invariant volume element (3.17). Thus, the way we take a functional determinant will depend on the choice of $\ell(\phi)$. This leads us to define

$$\overline{\det}(M_{xy}) \equiv \exp\left[i \int d^D x \sqrt{-\bar{g}} \ln(M)_{xx} \right]. \tag{3.22}$$

Finally, the functional integral element for a field Φ^I is $\overline{\mathcal{D}}\Phi^I \equiv \prod_x d\Phi^I(x)$, which involves the product of integral elements at every point in spacetime. How this infinite product is performed depends on the choice of spacetime metric and hence ℓ.

We can see this dependence explicitly using the identity $\prod_i A_i = \exp\left(\sum_i \ln(A_i)\right)$. This identity holds for discrete products, but can be readily generalised to continuous products, giving us

$$\overline{\mathcal{D}}\Phi^I = \exp\left[\int d^D x \sqrt{-\bar{g}(x)}\, \ln\left(d\Phi^I(x)\right) \right]. \tag{3.23}$$

We have written all quantities that depend non-trivially on the choice of ℓ with a bar $\bar{}$ to emphasise this dependence.

3.3 The Field and Configuration Spaces for Gravity

Now that we have addressed the cosmological frame problem, let us proceed to define the field and configuration spaces for (3.1) explicitly. Starting with the field space, we shall define the following set of coordinates on the field-space manifold

$$\Phi^I = \begin{pmatrix} g^{\mu\nu} \\ \phi^A \end{pmatrix}, \tag{3.24}$$

which we shall refer to collectively by $\mathbf{\Phi}$. Here, we have introduced the index $I \in \{(\mu\nu), A\}$, which runs from 1 to $N + D(D+1)/2$. In the gravity sector, the single index I represents a symmetrised *pair* of spacetime indices $(\mu\nu)$ and, thus, all gravitational degrees of freedom are included in (3.24).

Notice that we have chosen the coordinate of our field space to be the inverse metric $g^{\mu\nu}$, and not $g_{\mu\nu}$. This is purely conventional and is done for aesthetic reasons, since it allows an upper field space index to correspond to an upper spacetime index.

We see that, as before, field reparametrisations can be interpreted as diffeomorphisms of the field space. Indeed, (3.2) can be rewritten as

$$\Phi^I \to \widetilde{\Phi}^I(\mathbf{\Phi}). \tag{3.25}$$

As we mentioned previously, it is common to only consider the restricted set of redefinitions (3.3) and (3.4).

There is no need to make this restriction in our formalism. We note, however, that if we consider a transformation that is not of the form the form (3.3) or (3.4) then the relation (3.14) will no longer hold. Nevertheless, as long as we insist that $\bar{g}_{\mu\nu}$ continues to transform as a field space scalar, the spacetime line element (3.16) will not be affected by this transformation. Therefore, our formalism can be used to maintain manifest invariance under *any* transformation of the form (3.25).

In order to utilise the techniques of field space covariance we must equip the field space with a metric. Taking inspiration from (2.41) we, therefore, define

$$G_{IJ} = \frac{\bar{g}_{\mu\nu}}{4} \frac{\partial^2 \bar{\mathcal{L}}}{\partial(\partial_\mu \Phi^I)\partial(\partial_\nu \Phi^J)} - \frac{\partial^2 \bar{\mathcal{L}}}{\partial \Phi^I \partial\left(\overline{\Box}\phi^J\right)} - \frac{\partial^2 \bar{\mathcal{L}}}{\partial \Phi^I \partial\left(\overline{\Box}\phi^J\right)}, \tag{3.26}$$

where $\overline{\Box} \equiv \bar{g}^{\mu\nu}\nabla_\mu\nabla_\nu$ and we have defined

$$\bar{\mathcal{L}} = \ell^D \mathcal{L} \tag{3.27}$$

so that

$$S = \int d^D x \sqrt{-\bar{g}} \bar{\mathcal{L}}. \tag{3.28}$$

The Lagrangian $\bar{\mathcal{L}}$ defined in this way is a field space scalar, as opposed to $\mathcal{L}$, which picks up a conformal factor under (3.3).

Before we proceed, we need to address the issue of gauge invariance. The theory (3.1) is invariant under the transformation

$$g_{\mu\nu} \to g_{\mu\nu} + \partial_\mu \xi_\nu + \partial_\nu \xi_\mu \tag{3.29}$$

for any spacetime vector $\xi_\mu(\boldsymbol{x})$. This gauge invariance forces us to add to the action a gauge fixing term of the form

$$S_{\mathrm{GF}} = -\frac{1}{2} \int d^D x \, \gamma \sqrt{-\bar{g}} \chi^\mu \bar{g}_{\mu\nu} \chi^\nu. \tag{3.30}$$

Here, $\chi^\mu(g_{\rho\sigma}) = 0$ is the gauge fixing condition and γ is an arbitrary constant. Additionally, we must insert the Faddeev–Poppov determinant [64]

$$V_{\mathrm{FP}} = \overline{\det} \left(\frac{\bar{\delta} \chi^\mu(\boldsymbol{x})}{\bar{\delta} \xi^\nu(\boldsymbol{y})} \right) \tag{3.31}$$

into our path integral measure.

In this work we shall consider the De Donder gauge [65]

$$\chi^\mu = g^{\rho\sigma} \Gamma^\mu_{\rho\sigma} = 0. \tag{3.32}$$

We must note that (3.32) is a frame dependent statement valid only in one particular frame. In a general frame, χ^μ will depend on both the tensor field $g_{\mu\nu}$ and the scalar fields ϕ.

One may worry that the arbitrary choice of gauge fixing condition will affect the uniqueness of our approach. This is a genuine concern and one that has been much studied in the literature [4, 5, 8, 9, 66–69]. Indeed, one of the original motivations for construction of the VDW effective action was the gauge dependence of the effective action (2.88) shown by Jackiw [70].

These previous works have shown that one can obtain an expression for the effective action that is reparametrisation invariant, gauge invariant and invariant under the choice of gauge fixing condition. This was achieved by projecting the field space onto the space of gauge orbits so that the gauge direction (3.29) corresponds to zero distance in field space.

However, such considerations are not necessary for our purposes since we are only interested in reparametrisation invariance and not gauge invariance. The discussion of gauge freedom and the efforts to make the formalism gauge invariant run parallel to our goal of manifest reparametrisation invariance and we choose to set them aside for the time being in order to reduce complexity. We shall, therefore, not project out the gauge degrees of freedom and shall instead continue to work with the full set of $N + D(D+1)/2$ coordinates. Hence, we will proceed to use (3.26) in its unaltered form. We note that the techniques used to remove gauge dependence in these previous works can always be added to our formalism after the fact.

We can now calculate the field space metric for a scalar tensor theory by applying (3.26) to the sum of (3.1) and (3.30). We find

$$G_{IJ} = \ell^2 \begin{pmatrix} G_{(\mu\nu)(\rho\sigma)} & \frac{1-D}{D} g_{\rho\sigma} \partial_A f \\ \frac{1-D}{D} g_{\mu\nu} \partial_B f & k_{AB} \end{pmatrix} \tag{3.33}$$

where

$$G_{(\mu\nu)(\rho\sigma)} = \left(\frac{2-D}{2D} f - \frac{2\gamma}{D\ell^{D+2}} \right) (g_{\mu\rho} g_{\nu\sigma} + g_{\mu\sigma} g_{\nu\rho}) \tag{3.34}$$

$$+ \left(\frac{4-D}{D\ell^{D+2}} \gamma + \frac{D-2}{D} f \right) g_{\mu\nu} g_{\rho\sigma}. \tag{3.35}$$

We notice that the choice of gauge fixing term and the choice of normalisation constant γ both affect the form of the field space metric. We refer the reader to the previously discussed works [4, 5, 8, 9, 66–69] for efforts to remove the dependence on χ^μ and shall present an argument for how to fix γ in Sect. 3.4. For now though we shall keep the full expression and note that the dependence on the gauge fixing term does not affect the transformation properties of G_{IJ} or the frame invariance of our formalism.

The field space metric allows us to define field space connections

$$\Gamma^I_{JK} = \frac{1}{2} G^{IL} \left[\partial_J G_{LK} + \partial_K G_{JL} - \partial_L G_{JK} \right], \tag{3.36}$$

where $\partial_I \equiv \partial / \partial \Phi^I$. Hence, we can also define a field-space covariant derivative

$$\nabla_J X^I = \partial_I \Phi^J + \Gamma^I_{JK} X^K, \qquad \nabla_J X_I = \partial_I \Phi^J - \Gamma^K_{JI} X_K, \tag{3.37}$$

with straightforward generalisation to higher order tensors.

As in the scalar case, it is desirable to generalise the above construction to the infinite-dimensional configuration space. This is straightforward to achieve. We start by generalising the coordinates to

$$\Phi^{\bar{I}} = \Phi^I(x_I). \tag{3.38}$$

As mentioned before we are using indices with a bar instead of a hat to emphasise that the definition of these coordinates (and, in particular their contraction) depends on the spacetime metric $\bar{g}_{\mu\nu}$.

We can also generalise (2.61) to equip the configuration space with a metric

$$\mathcal{G}_{\bar{I}\bar{J}} = \frac{\bar{g}_{\mu\nu}}{D} \frac{\bar{\delta}^2 S}{\bar{\delta}(\partial_\mu \Phi^{\bar{I}})\bar{\delta}(\partial_\nu \Phi^{\bar{J}})} = G_{IJ}(x_I)\bar{\delta}^{(D)}(x_I - x_J). \tag{3.39}$$

This leads to the configuration space line element

$$\mathcal{D}\Sigma^2[\mathbf{\Phi}] = \mathcal{G}_{\bar{I}\bar{J}}\mathcal{D}\Phi^{\bar{I}}\mathcal{D}\Phi^{\bar{J}} = \int d^D x \sqrt{-\bar{g}}\, G_{IJ}(x)\mathcal{D}\Phi^I(x)\mathcal{D}\Phi^J(x)\,. \tag{3.40}$$

The definition of the configuration space metric leads to the following configuration space connections

$$\begin{aligned}\Gamma^{\bar{I}}_{\bar{J}\bar{K}} &= \frac{1}{2}\mathcal{G}^{\bar{I}\bar{L}}\left[\frac{\bar{\delta}\mathcal{G}_{\bar{J}\bar{L}}}{\bar{\delta}\Phi^{\bar{K}}} + \frac{\bar{\delta}\mathcal{G}_{\bar{L}\bar{K}}}{\bar{\delta}\Phi^{\bar{J}}} - \frac{\bar{\delta}\mathcal{G}_{\bar{J}\bar{K}}}{\bar{\delta}\Phi^{\bar{L}}}\right] \\ &= \left[\Gamma^I_{JK} - \frac{1}{4\bar{g}}\left(\delta^I_J\partial_K\bar{g} + \delta^I_K\partial_J\bar{g} - G_{JK}\partial^I\bar{g}\right)\right]\bar{\delta}^{(D)}(x_I - x_J)\bar{\delta}^{(D)}(x_I - x_J),\end{aligned} \tag{3.41}$$

and, hence, the configuration-space covariant derivative

$$\overline{\nabla}_{\bar{J}}X^{\bar{I}} = \frac{\bar{\delta}X^{\bar{I}}}{\bar{\delta}\Phi^{\bar{J}}} + \Gamma^{\bar{I}}_{\bar{J}\bar{K}}X^{\bar{K}}, \qquad \overline{\nabla}_{\bar{J}}X_{\bar{I}} = \frac{\bar{\delta}X_{\bar{I}}}{\bar{\delta}\Phi^{\bar{J}}} - \Gamma^{\bar{K}}_{\bar{J}\bar{I}}X_{\bar{K}}, \tag{3.42}$$

with, again, straightforward generalisation to higher order tensors.

Notice that

$$\Gamma^{\bar{I}}_{\bar{J}\bar{K}} \neq \Gamma^I_{JK}\bar{\delta}^{(D)}(x_I - x_J)\bar{\delta}^{(D)}(x_I - x_J) \tag{3.43}$$

in contrast to the scalar case. This is because $\bar{\delta}^{(D)}(\boldsymbol{x} - \boldsymbol{y})$ has non trivial dependencies on the fields through the spacetime metric. Indeed

$$\partial_I\, \bar{\delta}^{(D)}(\boldsymbol{x} - \boldsymbol{y}) = -\frac{1}{2}\frac{\partial_I\bar{g}}{\bar{g}}\,\bar{\delta}^{(D)}(\boldsymbol{x} - \boldsymbol{y}) \neq 0. \tag{3.44}$$

Note that such a dependence is a generic consequence of treating $g_{\mu\nu}$ as a field and would have arisen even if we had not introduced the model function ℓ.

Finally, we see that the configuration space comes equipped with a reparametrisation-invariant volume element

$$\left[\overline{\mathcal{D}}\mathcal{V}\right] = \left[\overline{\mathcal{D}}\mathbf{\Phi}\right]\sqrt{\det\left(\mathcal{G}_{\bar{I}\bar{J}}\right)}, \tag{3.45}$$

where

$$\left[\overline{\mathcal{D}}\mathbf{\Phi}\right] = \prod_I \overline{\mathcal{D}}\Phi^I. \tag{3.46}$$

This volume element will be needed in order to define path integrals in a reparametrisation invariant way.

3.4 Example: Einstein Hilbert Action

As an example, let us consider standard General Relativity in D dimensions as described by the Einstein Hilbert action [71]

$$S = -\frac{1}{2}\int d^D x\sqrt{-g}\,[R + \Lambda], \tag{3.47}$$

where Λ is the cosmological constant. This is of the form (3.1), but the absence of scalar fields in this theory forces us to choose

$$\ell = 1, \qquad f = 1, \qquad k_{AB} = 0, \qquad V = \Lambda. \tag{3.48}$$

We can define the field space for this theory as a $D(D+1)/2$ dimensional manifold with coordinates

$$\Phi^{(\mu\nu)} = g^{\mu\nu}. \tag{3.49}$$

As before, each symmetrised pair of spacetime indices $(\mu\nu)$ represents a single field space index. In this section, however, we shall write out both spacetime indices explicitly for clarity.

This theory still retains the gauge symmetry (3.29) and we must, therefore, add a gauge fixing term of the form (3.30) as before. In this section we shall keep the gauge fixing term general and note that, whatever we choose for χ^μ and γ, the field space metric must be of the form

$$G_{(\mu\nu)\,(\rho\sigma)} = -\frac{\mathcal{N}}{4}\left(\alpha g_{\mu\rho}g_{\sigma\nu} + \alpha g_{\mu\sigma}g_{\rho\nu} - 2g_{\mu\nu}g_{\rho\sigma}\right) \tag{3.50}$$

due to its symmetries and spacetime properties. Here, $\mathcal{N}$ is an irrelevant normalisation and $\alpha(\gamma, \chi^\mu)$ is a constant that depends on both the gauge fixing condition χ^μ and the constant γ. For example, in the De Donder gauge (3.32) in 4-dimensions we have $\alpha = 1 + 2\gamma$.

Because the gauge fixing term is, in principle, completely arbitrary we need a new condition to fix the value of α. The condition we choose is

$$\begin{aligned}\left(G^{-1}\right)^{(\mu\nu)(\rho\sigma)} &= G^{(\mu\nu)(\rho\sigma)} \\ &\equiv g^{\alpha\mu}g^{\beta\nu}g^{\kappa\rho}g^{\lambda\sigma}\,G_{(\alpha\beta)\,(\kappa\lambda)},\end{aligned} \tag{3.51}$$

where $\left(G^{-1}\right)^{(\mu\nu)(\rho\sigma)}$ is the inverse metric satisfying

$$G_{(\mu\nu)(\rho\sigma)}\left(G^{-1}\right)^{(\rho\sigma)\,(\kappa\lambda)} = \frac{1}{2}(\delta^\mu_\rho\delta^\nu_\sigma + \delta^\mu_\sigma\delta^\nu_\rho). \tag{3.52}$$

This is a useful condition to satisfy because it means there is no difference between raising indices with the spacetime metric $\bar{g}^{\mu\nu}$ or the field space metric $\left(G^{-1}\right)^{(\mu\nu)(\rho\sigma)}$.

We can calculate the matrix inverse of (3.50) and we find it to be

$$\left(G^{-1}\right)^{(\mu\nu)(\rho\sigma)} = -\frac{1}{\mathcal{N}\alpha}\left(g^{\mu\rho}g^{\nu\sigma} + g^{\mu\sigma}g^{\nu\rho} - \frac{2}{D-\alpha}g^{\mu\nu}g^{\rho\sigma}\right). \tag{3.53}$$

We, therefore, see that the solution to (3.51) is

$$\mathcal{N} = \frac{4}{D}, \qquad \alpha = \frac{D}{2} \tag{3.54}$$

and thus

$$G_{(\mu\nu)(\rho\sigma)} = \frac{1}{2}\left(g_{\mu\rho}g_{\sigma\nu} + g_{\mu\sigma}g_{\rho\nu} - \frac{4}{D}g_{\mu\nu}g_{\rho\sigma}\right). \tag{3.55}$$

This result agrees with the metric calculated in four dimensions by Vilkovisky in [8, 9] using different considerations as well as other results in the literature [72, 73].[4]

Let us analyse the field space of gravity by calculating the curvature invariants of (3.55). In order to deal with the complexity of the calculations involved we have employed the symbolic computer algebra package `Cadabra2` [74, 75].

We start by calculating the field space Christoffel symbols, which are

$$\begin{aligned}\Gamma^{(\alpha\beta)}{}_{(\mu\nu)\,(\rho\sigma)} = \frac{1}{4}\Big(&g_{\mu\rho}\delta^{\alpha}_{\nu}\delta^{\beta}_{\sigma} + g_{\mu\sigma}\delta^{\alpha}_{\sigma}\delta^{\beta}_{\nu} + g_{\mu\sigma}\delta^{\alpha}_{\nu}\delta^{\beta}_{\rho} + g_{\mu\sigma}\delta^{\alpha}_{\rho}\delta^{\beta}_{\nu} \\ &+ g_{\nu\rho}\delta^{\alpha}_{\mu}\delta^{\beta}_{\sigma} + g_{\nu\sigma}\delta^{\alpha}_{\sigma}\delta^{\beta}_{\mu} + g_{\nu\sigma}\delta^{\alpha}_{\mu}\delta^{\beta}_{\rho} + g_{\nu\sigma}\delta^{\alpha}_{\rho}\delta^{\beta}_{\mu}\Big).\end{aligned} \tag{3.56}$$

We can then calculate the Riemann tensor, which we find to be

$$\begin{aligned}&\mathfrak{R}^{(\mu\nu)}{}_{(\alpha\beta)(\rho\sigma)(\gamma\delta)} \\ &\equiv \partial_{(\rho\sigma)}\Gamma^{(\mu\nu)}_{(\gamma\delta)(\alpha\beta)} - \partial_{(\gamma\delta)}\Gamma^{(\mu\nu)}_{(\rho\sigma)(\alpha\beta)} + \Gamma^{(\mu\nu)}_{(\rho\sigma)(\kappa\lambda)}\Gamma^{(\kappa\lambda)}_{(\gamma\delta)(\alpha\beta)} - \Gamma^{(\mu\nu)}_{(\gamma\delta)(\kappa\lambda)}\Gamma^{(\kappa\lambda)}_{(\rho\sigma)(\alpha\beta)} \\ &= -\frac{1}{32}\delta^{\mu}_{\rho}\delta^{\nu}_{\beta}g_{\sigma\gamma}g_{\alpha\delta} + \frac{1}{32}\delta^{\mu}_{\gamma}\delta^{\nu}_{\beta}g_{\rho\delta}g_{\sigma\alpha} + \frac{1}{4D}g_{\rho\gamma}g^{\mu\nu}g_{\sigma\alpha}g_{\beta\delta} - \frac{1}{4D}g^{\mu\nu}g_{\rho\beta}g_{\sigma\gamma}g_{\alpha\delta} \\ &+ (\alpha\leftrightarrow\beta) + (\mu\leftrightarrow\nu) + (\rho\leftrightarrow\sigma) + (\gamma\leftrightarrow\delta).\end{aligned} \tag{3.57}$$

We can also contract the Riemann tensor to calculate the Ricci tensor,

[4] Note that (3.55) is not the DeWitt metric [12]. The DeWitt metric imposes a time slicing condition and focuses only on the spatial components of $g_{\mu\nu}$. In contrast, our approach treats all components on an equal footing and this allows our metric to transform as a spacetime tensor.

$$\begin{aligned}\mathfrak{R}_{(\mu\nu)(\rho\sigma)} &= \mathfrak{R}^{(\gamma\delta)}{}_{(\mu\nu)(\gamma\delta)(\rho\sigma)},\\ &= \frac{1}{4}g_{\mu\nu}g_{\rho\sigma} - \frac{D}{8}g_{\mu\rho}g_{\nu\sigma} - \frac{D}{8}g_{\mu\sigma}g_{\nu\rho},\end{aligned} \tag{3.58}$$

and we can contract the Ricci tensor to compute the Ricci scalar

$$\begin{aligned}\mathfrak{R} &= \left(G^{-1}\right)^{(\alpha\beta)\,(\gamma\delta)}\,\mathfrak{R}_{(\alpha\beta)(\gamma\delta)}\\ &= \frac{D}{4} - \frac{D^2}{8} - \frac{D^3}{8}.\end{aligned} \tag{3.59}$$

Notice that all of these curvature invariants are non-zero except for the trivial case $D = 1$.[5] This means that the field space for gravity is curved. Indeed, from (3.59) we see that the field space is negatively curved in all dimensions $D > 1$. It would be interesting to see whether the non-renormalisability of (3.47) is related to this negative curvature.

3.5 Unique Frame Invariant Effective Action for Quantum Gravity

To finish this chapter let us present the unique frame invariant effective action for scalar-tensor theories with action (3.1). As we have stated previously, to fully define a scalar tensor theory one must define the following four model functions:

1. The effective Planck length ℓ.
2. The effective Planck mass f.
3. The scalar field-space metric k_{AB}.
4. The scalar potential V.

With these defined it is possible to construct the field and configuration spaces as in Sect. 3.3 and, hence, we can define the effective action in a way that is unique, frame invariant and spacetime diffeomorphism invariant.

We use the work of Vilkovisky and DeWitt discussed in Chap. 2 to define

$$\Sigma^{\bar{I}}[\varphi, \mathbf{\Phi}] = (C^{-1}[\varphi])^{\bar{I}}{}_{\bar{J}}\,\sigma^{\bar{J}}[\varphi, \mathbf{\Phi}] \tag{3.60}$$

where

$$\sigma^{\bar{I}}[\varphi, \mathbf{\Phi}] = (\Phi^{\bar{I}} - \varphi^{\bar{I}}) + \frac{1}{2}\Gamma^{\bar{I}}_{\bar{J}\bar{K}}[\varphi](\Phi^{\bar{J}} - \varphi^{\bar{J}})(\Phi^{\bar{K}} - \varphi^{\bar{K}}) + \cdots . \tag{3.61}$$

is the tangent to the geodesic linking φ and $\mathbf{\Phi}$ and

[5] It is impossible to have curvature in one dimension so the fact that the curvature invariants vanish in this case is fully expected.

$$C^{\bar{I}}{}_{\bar{J}}[\varphi] = -\langle \overline{\nabla}_{\bar{J}} \sigma^{\bar{I}}[\varphi, \boldsymbol{\Phi}] \rangle \approx \langle \delta^{\bar{I}}_{\bar{J}} - \frac{1}{3} R^{\bar{I}}{}_{\bar{K}\bar{J}\bar{L}}[\varphi]\, \sigma^{\bar{K}}[\varphi, \boldsymbol{\Phi}]\, \sigma^{\bar{L}}[\varphi, \boldsymbol{\Phi}] + \ldots \rangle \,. \tag{3.62}$$

We can, therefore, define the effective action

$$\exp\left(\frac{i}{\hbar}\Gamma[\boldsymbol{\Phi}]\right) = \int [\overline{\mathcal{D}}\varphi] \sqrt{\overline{\det}\,(\mathcal{G}_{\bar{I}\bar{J}})}\; V_{\mathrm{FP}} \exp\left[\frac{i}{\hbar}\left(S[\varphi] - \frac{\bar{\delta}\Gamma}{\bar{\delta}\Phi^{\bar{I}}} \Sigma^{\bar{I}}[\varphi, \boldsymbol{\Phi}]\right)\right]. \tag{3.63}$$

The effective action defined in (3.63) is frame and spacetime diffeomorphism invariant as required. Furthermore, it reduces to previous, frame-dependent, definitions in the literature [4, 5, 8, 9, 52, 53, 66–69] in the metric frame when $\ell = 1$. Additionally it reduces to the standard effective action when calculations are performed on shell and so agrees with all observations.

We note that the measure (3.45) has a non-trivial dependence on $\ell(\phi)$ even when the calculations are performed on shell. Therefore, theories with different values for this model function will pick up different quantum corrections even if the classical action S is identical. This difference is equivalent to the effects found by Falls and Herrero-Valea [6]. However, instead of interpreting this as a frame dependence, we see that the "frame discriminant" is nothing but the quantum effects of a non-trivial $\ell(\phi)$.

References

1. Dicke RH (1962) Mach's principle and invariance under transformation of units. Phys Rev 125:2163–2167
2. Burns D, Karamitsos S, Pilaftsis A (2016) Frame-covariant formulation of inflation in scalar-curvature theories. Nucl Phys B907:785–819. arXiv: 1603.03730 [hep-ph] (cit. on pp. 23, 80, 83)
3. Karamitsos S, Pilaftsis A (2018) Frame covariant nonminimal multifield inflation. Nucl Phys B927:219–254. arXiv: 1706.07011 [hep-ph] (cit. on pp. 23, 80, 81, 83)
4. Kunstatter G (1986) Vilkovisky's unique effective action: an introduction and Explicit Calculation". NATO Advanced Research Workshop on Superfield Theories Vancouver, British Columbia, Canada, July 25-August 6, 1986. 1986, pp. 503–517 (cit. on pp. 30, 51, 83, 90, 91, 96)
5. G. Kunstatter and H.P. Leivo. "On the Reparametrization Dependence of the Effective Potential". In: Phys. Lett. B 183 (1987), pp. 75–80 (cit. on pp. 30, 51, 83, 90, 91, 96)
6. Kevin Falls and Mario Herrero-Valea. "Frame (In)equivalence in Quantum Field Theory and Cosmology". In: Eur. Phys. J. C 79.7 (2019), p. 595. arXiv: 1812.08187 [hep-th] (cit. on pp. 30, 83, 84, 86, 97, 245)
7. Sotirios Karamitsos and Apostolos Pilaftsis. "On the Cosmological Frame Problem". In: PoS CORFU2017 (2018), p. 036. arXiv: 1801.07151 [hep-th] (cit. on pp. 31, 83)
8. G. A. Vilkovisky. "The Unique Effective Action in Quantum Field Theory". In: Nucl. Phys. B234 (1984), pp. 125-137 (cit. on pp. 50, 51, 55, 56, 63, 64, 67-69, 79, 90, 91, 94, 96, 124, 243)
9. G. A. Vilkovisky. "The Gospel According to DeWitt". 1984, pp. 169–209 (cit. on pp. 50, 51, 55, 56, 63, 64, 67-69, 79, 90, 91, 94, 96, 124, 243)

10. Bryce S. DeWitt. "Quantum Theory of Gravity. 2. The Manifestly Covariant Theory". In: Phys. Rev. 162 (1967). [,298(1967)], pp. 1195–1239 (cit. on pp. 55, 79)
11. Bryce S. DeWitt. "Quantum Theory of Gravity. 3. Applications of the Covariant Theory". In: Phys. Rev. 162 (1967). [,307(1967)], pp. 1239–1256 (cit. on p. 79)
12. Bryce S. DeWitt. "Quantum Theory of Gravity. 1. The Canonical Theory". In: Phys. Rev. 160 (1967). [3,93(1987)], pp. 1113–1148 (cit. on pp. 79, 94)
13. Kazuo Fujikawa. "Path Integral Measure for Gravitational Interactions". In: Nucl. Phys. B226 (1983), pp. 437–443 (cit. on p. 79)
14. G. Kunstatter. "Geometrical approach to the effective action". Banff Summer Institute: Gravitation. Oct. 1990, pp. 356–400 (cit. on pp. 79, 83)
15. Ian G. Moss. "Covariant one-loop quantum gravity and Higgs inflation" (Sept. 2014). arXiv: 1409.2108 [hep-th] (cit. on p. 79)
16. P. Jordan. "Zur empirischen Kosmologie". In: Naturwissenschaften 26.26 (1938), pp. 417–421 (cit. on p. 79)
17. P. Jordan. Schwerkraft und Weltall: Die Wissenschaft, Bd. 107. Braunschweig, 1952 (cit. on p. 79)
18. C. Brans and R. H. Dicke. "Mach's principle and a relativistic theory of gravitation". In: Phys. Rev. 124 (1961), pp. 925–935 (cit. on pp. 79, 81)
19. Peter G. Bergmann. "Comments on the scalar tensor theory". In: Int. J. Theor. Phys. 1 (1968), pp. 25–36 (cit. on p. 79)
20. Robert V. Wagoner. "Scalar tensor theory and gravitational waves". In: Phys. Rev. D1 (1970), pp. 3209–3216 (cit. on p. 79)
21. Y. Fujii and K. Maeda. The scalar-tensor theory of gravitation. Cambridge Monographs on Mathematical Physics. Cambridge University Press, 2007 (cit. on p. 79)
22. Valerio Faraoni. Cosmology in scalar tensor gravity. Vol. 139. 2004 (cit. on p. 79)
23. Sotirios Karamitsos. "Beyond the Poles in Attractor Models of Inflation". In: JCAP 09 (2019), p. 022. arXiv: 1903.03707 [hep-th] (cit. on p. 80)
24. Gerard 't Hooft and M.J.G. Veltman. "One loop divergencies in the theory of gravitation". In: Ann. Inst. H. Poincare Phys. Theor. A 20 (1974), pp. 69–94 (cit. on p. 80)
25. Marc H. Goroff and Augusto Sagnotti. "Quantum Gravity at Two Loops". In: Phys. Lett. B 160 (1985), pp. 81–86 (cit. on p. 80)
26. V. Faraoni and E. Gunzig. "Einstein frame or Jordan frame?" In: Int. J. Theor. Phys. 38 (1999), pp. 217–225. arXiv: astro-ph/9910176 (cit. on p. 81)
27. Y.M. Cho. "Reinterpretation of Jordan-Brans-Dicke theory and Kaluza-Klein cosmology". In: Phys. Rev. Lett. 68 (1992), pp. 3133–3136 (cit. on p. 82)
28. A. Friedmann. "Über die Krümmung des Raumes". In: Zeitschrift für Physik 10.1 (1922), pp. 377–386 (cit. on pp. 82, 173)
29. A. Friedmann. "Über die Möglichkeit einer Welt mit konstanter negative Krümmung des Raumes". In: Zeitschrift für Physik 21.1 (1924), pp. 326–332 (cit. on pp. 82, 173)
30. Abbé G. Lemaître. "A Homogeneous Universe of Constant Mass and Increasing Radius accounting for the Radial Velocity of Extra-galactic Nebulæ". In: Monthly Notices of the Royal Astronomical Society 91.5 (Mar. 1931), pp. 483–490 (cit. on pp. 82, 173)
31. H. P. Robertson. "Kinematics and World-Structure". In: Astrophys. J. 82 (Nov. 1935), p. 284 (cit. on pp. 82, 173)
32. H. P. Robertson. "Kinematics and World-Structure II." In: Astrophys. J. 83 (Apr. 1936), p. 187 (cit. on pp. 82, 173)
33. H. P. Robertson. "Kinematics and World-Structure III." In: Astrophys. J. 83 (May 1936), p. 257 (cit. on pp. 82, 173)
34. A. G. Walker. "On Milne's Theory of World-Structure*". In: Proceedings of the London Mathematical Society s2-42.1 (1937), pp. 90–127 (cit. on pp. 82, 173)
35. Ignatios Antoniadis, C. Bachas, John R. Ellis, and Dimitri V. Nanopoulos. "An Expanding Universe in String Theory". In: Nucl. Phys. B 328 (1989), pp. 117–139 (cit. on p. 82)
36. S. Capozziello, R. de Ritis, and Alma Angela Marino. "Some aspects of the cosmological conformal equivalence between 'Jordan frame' and 'Einstein frame'". In: Class. Quant. Grav. 14 (1997), pp. 3243–3258. arXiv: gr - qc/9612053 [gr-qc] (cit. on p. 82)

37. Valerio Faraoni, Edgard Gunzig, and Pasquale Nardone. "Conformal transformations in classical gravitational theories and in cosmology". In: Fund. Cosmic Phys. 20 (1999), p. 121. arXiv: gr-qc/9811047 [gr-qc] (cit. on p. 82)
38. Enrique Alvarez and Jorge Conde. "Are the string and Einstein frames equivalent". In: Mod. Phys. Lett. A17 (2002), pp. 413–420. arXiv: gr - qc/0111031 [gr-qc] (cit. on p. 82)
39. S. Capozziello, P. Martin-Moruno, and C. Rubano. "Physical non-equivalence of the Jordan and Einstein frames". In: Phys. Lett. B689 (2010), pp. 117–121. arXiv: 1003.5394 [gr-qc] (cit. on p. 82)
40. Y.M. Cho and D.H. Park. "Higher dimensional unification and fifth force". In: Nuovo Cim. B 105 (1990), pp. 817–829 (cit. on p. 82)
41. Y.M. Cho and D.H. Park. "Fifth force from Kaluza-Klein unification". In: Gen. Rel. Grav. 23 (1991), pp. 741–757 (cit. on p. 82)
42. Guillem Domènech and Misao Sasaki. "Conformal Frame Dependence of Inflation". In: JCAP 1504.04 (2015), p. 022. arXiv: 1501.07699 [gr-qc] (cit. on p. 82)
43. Debottam Nandi and Pankaj Saha. "Einstein or Jordan: seeking answers from the reheating constraints" (2019). arXiv: 1907.10295 [gr-qc] (cit. on p. 82)
44. Marieke Postma and Marco Volponi. "Equivalence of the Einstein and Jordan frames". In: Phys. Rev. D90.10 (2014), p. 103516. arXiv: 1407.6874 [astro-ph.CO] (cit. on p. 83)
45. Laur Järv, Kristjan Kannike, Luca Marzola, et al. "Frame-Independent Classification of Single-Field Inflationary Models". In: Phys. Rev. Lett. 118.15 (2017), p. 151302. arXiv: 1612.06863 [hep-ph] (cit. on p. 83)
46. Sachin Pandey and Narayan Banerjee. "Equivalence of Jordan and Einstein frames at the quantum level". In: Eur. Phys. J. Plus 132.3 (2017), p. 107. arXiv: 1610.00584 [gr-qc] (cit. on p. 83)
47. Jérémie Francfort, Basundhara Ghosh, and Ruth Durrer. "Cosmological Number Counts in Einstein and Jordan frames". In: JCAP 09 (2019), p. 071. arXiv: 1907.03606 [gr-qc] (cit. on p. 83)
48. A.O. Barvinsky and G.A. Vilkovisky. "The Generalized Schwinger-Dewitt Technique in Gauge Theories and Quantum Gravity". In: Phys. Rept. 119 (1985), pp. 1–74 (cit. on p. 83)
49. Christian F. Steinwachs and Alexander Yu. Kamenshchik. "One-loop divergences for gravity non-minimally coupled to a multiplet of scalar fields: calculation in the Jordan frame. I. The main results". In: Phys. Rev. D 84 (2011), p. 024026. arXiv: 1101.5047 [gr-qc] (cit. on p. 83)
50. Christian F. Steinwachs and Alexander Yu. Kamenshchik. "Non-minimal Higgs Inflation and Frame Dependence in Cosmology". In: AIP Conf. Proc. 1514.1 (2013), pp. 161–164. arXiv: 1301.5543 [gr-qc] (cit. on p. 83)
51. Nobuyoshi Ohta. "Quantum equivalence of f(R) gravity and scalar-tensor theories in the Jordan and Einstein frames". In: PTEP 2018.3 (2018), 033B02. arXiv: 1712.05175 [hep-th] (cit. on p. 83)
52. Alexander Yu. Kamenshchik and Christian F. Steinwachs. "Question of quantum equivalence between Jordan frame and Einstein frame". In: Phys. Rev. D91.8 (2015), p. 084033. arXiv: 1408.5769 [gr-qc] (cit. on pp. 83,96)
53. Marios Bounakis and Ian G. Moss. "Gravitational corrections to Higgs potentials". In: JHEP 04 (2018), p. 071. arXiv: 1710.02987 [hep-th] (cit. on pp. 83, 96)
54. 3S. R. Huggins, G. Kunstatter, H. P. Leivo, and D. J. Toms. "On the Unique Effective Action in Five-dimensional Kaluza-Klein Theory". In: Phys. Rev. Lett. 58 (1987), p. 296 (cit. on p. 83)
55. S.R. Huggins, G. Kunstatter, H.P. Leivo, and D.J. Toms. "The Vilkovisky-de Witt Effective Action for Quantum Gravity". In: Nucl. Phys. B 301 (1988), pp. 627–660 (cit. on p. 83)
56. I.L. Buchbinder, P.M. Lavrov, and S.D. Odintsov. "Unique Effective Action in Kaluza-Klein Quantum Theories and Spontaneous Compactification". In: Nucl. Phys. B 308 (1988), pp. 191–202 (cit. on p. 83)
57. Theodor Kaluza. "Zum Unitätsproblem der Physik". In: Sitzungsber. Preuss. Akad. Wiss. Berlin (Math. Phys.) 1921 (1921), pp. 966–972. arXiv: 1803. 08616 [physics.hist-ph] (cit. on pp. 83, 139)

58. Oskar Klein. "Quantum Theory and Five-Dimensional Theory of Relativity. (In German and English)". In: Z. Phys. 37 (1926). Ed. by J.C. Taylor, pp. 895–906 (cit. on pp. 83, 139)
59. P. W. Higgs. "Quadratic lagrangians and general relativity". In: Nuovo Cim. 11.6 (1959), pp. 816–820 (cit. on p. 85)
60. Eanna E. Flanagan. "The Conformal frame freedom in theories of gravitation". In: Class. Quant. Grav. 21 (2004), p. 3817. arXiv: gr-qc/0403063 [gr-qc] (cit. on p. 85)
61. Riccardo Catena, Massimo Pietroni, and Luca Scarabello. "Einstein and Jordan reconciled: a frame-invariant approach to scalar-tensor cosmology". In: Phys. Rev. D76 (2007), p. 084039. arXiv: astro-ph/0604492 [astro-ph] (cit. on p. 85)
62. Laur Järv, Piret Kuusk, Margus Saal, and Ott Vilson. "Invariant quantities in the scalar-tensor theories of gravitation". In: Phys. Rev. D91.2 (2015), p. 024041. arXiv: 1411.1947 [gr-qc] (cit. on p. 85)
63. Alexandros Karam, Angelos Lykkas, and Kyriakos Tamvakis. "Frame-invariant approach to higher-dimensional scalar-tensor gravity". In: Phys. Rev. D97.12 (2018), p. 124036. arXiv: 1803.04960 [gr-qc] (cit. on p. 85)
64. L.D. Faddeev and V.N. Popov. "Feynman Diagrams for the Yang-Mills Field". In: Phys. Lett. B 25 (1967). Ed. by Jong-Ping Hsu and D. Fine, pp. 29–30 (cit. on p. 89)
65. T. Donder. Applications de la gravifique einsteinienne. Applications de la gravifique einsteinienne no. 43. Gauthier-Villars et cie, 1930 (cit. on p. 89)
66. Bryce S. DeWitt. "A Gauge Invariant Effective Action". Oxford Conference on Quantum Gravity. July 1980, pp. 449–487 (cit. on pp. 90, 91, 96)
67. Bryce S. DeWitt. "The Effective Action". Quantum Field Theory and Quantum Statistics, Vol. 1. 1988, pp. 191-222 (cit. on pp. 90, 91, 96)
68. E.S. Fradkin and Arkady A. Tseytlin. "On the New Definition of Off-shell Effective Action". In: Nucl. Phys. B 234 (1984), pp. 509–523 (cit. on pp. 90, 91, 96)
69. S.D. Odintsov. "The Parametrization Invariant and Gauge Invariant Effective Actions in Quantum Field Theory". In: Fortsch. Phys. 38 (1990), pp. 371–391 (cit. on pp. 90, 91, 96)
70. R. Jackiw. "Functional evaluation of the effective potential". In: Phys. Rev. D 9 (1974), p. 1686 (cit. on p. 90)
71. David Hilbert. "Die Grundlagen der Physik". In: Mathematische Annalen 92.1 (1924), pp. 1–32 (cit. on p. 93)
72. S.D. Odintsov. "Does the Vilkovisky-De Witt effective action in quantum gravity depend on the configuration space metric?" In: Phys. Lett. B 262 (1991), pp. 394–397 (cit. on p. 94)
73. Domenico Giulini and Claus Kiefer. "Wheeler-DeWitt metric and the attractivity of gravity". In: Phys. Lett. A193 (1994), pp. 21-24. arXiv: grqc/ 9405040 [gr-qc] (cit. on p. 94)
74. Kasper Peeters. "Introducing Cadabra: A Symbolic computer algebra system for field theory problems" (2007). arXiv: hep-th/0701238 [hep-th] (cit. on p. 94)
75. Kasper Peeters. "Cadabra2: computer algebra for field theory revisited". In: Journal of Open Source Software 3 (Dec. 2018), p. 1118 (cit. on p. 94)

Chapter 4
Field Space Covariance for Fermionic Theories

Where there is matter, there is geometry.

—Johannes Kepler

Fermions have received less attention in the context of field space covariance. Although early work by DeWitt [3] and Rebhan [4] showed that the VDW formalism applied just as well to fermionic degrees of freedom, they were unable to explicitly construct the field space manifold and, in particular, made no attempt to define the field space metric for such theories.

More recently [5, 6] there have been some attempts to define a fermionic field space metric, but these have been specific to the models under consideration. A general construction of the field space for fermionic theories is, therefore, still missing.

In this chapter we will show how such a construction can be achieved. As we have seen, field space covariance is now well understood for theories with scalar, tensor and gauge fields. Thus, fermions represent the last brick in the wall. Extending the VDW formalism to include such degrees of freedom will allow us to describe phenomenologically relevant theories of particle physics, including the Standard Model, in a geometric, reparametrisation invariant way.

4.1 Grassmann Numbers and Supermanifolds

Fermionic fields anticommute with one another [7–9]. This means that we need to introduce new mathematics to describe them, even at the classical level. The mathematical objects required are known as Grassmann numbers [10]. Grassmann numbers are defined to be a set of objects, usually denoted by θ_i, that anticommute with one another. Therefore,

K. Finn, *Geometric Approaches to Quantum Field Theory*, Springer Theses,
https://doi.org/10.1007/978-3-030-85269-6_4

$$\theta_i\theta_j = -\theta_j\theta_i. \tag{4.1}$$

An immediate consequence is that all Grassmann numbers are nilpotent with

$$\theta_i^2 = 0. \tag{4.2}$$

Because of (4.2) the most general function of a Grassmann number θ is

$$F(\theta) = A + \theta B, \tag{4.3}$$

where A and B do not depend on θ. This leads to a straightforward definition of differentiation with respect to Grassmann numbers

$$\frac{\overrightarrow{\partial}}{\partial\theta}F(\theta) = B. \tag{4.4}$$

However, note that in (4.3) we have chosen to write θB with θ on the left. Because θ and B do not necessarily commute, if we instead write θ on the right we would find

$$F(\theta) = A + \widetilde{B}\theta \tag{4.5}$$

with $\widetilde{B} \neq B$ in general. This choice, therefore, leads to a distinction between the left derivative defined in (4.4) and the right derivative defined

$$F(\theta)\frac{\overleftarrow{\partial}}{\partial\theta} = \widetilde{B}. \tag{4.6}$$

We can also define integration with respect to Grassmann numbers. The most common definition is due to Berezin [11] in which integration is taken to be identical to right differentiation. Thus, we define

$$\int F(\theta)d\theta = \widetilde{B}. \tag{4.7}$$

Although this definition may appear counterintuitive at first, it satisfies several identities that we commonly use to manipulate standard integrals including

$$\int [aF(\theta) + bG(\theta)]\,d\theta = a\int F(\theta)d\theta + b\int G(\theta)d\theta, \tag{4.8}$$

$$\int F(\theta+\phi)d\theta = \int F(\theta)d\theta. \tag{4.9}$$

It also satisfies the the identity

$$\int \frac{\overrightarrow{\partial}}{\partial\theta} F(\theta) d\theta = \int F(\theta) \frac{\overleftarrow{\partial}}{\partial\theta} d\theta = 0, \tag{4.10}$$

which allows us to continue to use integration by parts. The Berezinian integral can also be readily generalised to a functional integral in order to extend the path integral formalism to fermionic theories [12–14].

However, as we have previously seen, standard path integral techniques can lead to results that depend on the parametrisation of the fields. In order to study this parametrisation dependence, and ultimately remove it, we wish to construct a field space for theories with fermionic degrees of freedom. This means generalising our concept of a manifold to allow coordinates to be described by Grassmann numbers. Such a generalisation is known as a *supermanifold* [11, 15–21].

The mathematics of supermanifolds was first developed in the context of Supersymmetry (SUSY) [22–26], from which it takes its name. In SUSY, the spacetime manifold is extended with one or more Grassmannian coordinates and diffeomorphisms of these coordinates yield new Grassmannian Noether symmetries (also known as supersymmetries) of the theory. However, the study of supermanifolds has since become a subject in its own right and now has applications in both mathematics and physics far beyond SUSY [27–29].

It is in this latter context that we employ supermanifolds here. We will not be extending the spacetime manifold, but the field space manifold. Therefore, the new Grassmannian diffeomorphisms will not correspond to new Noether supersymmetries, but to reparametrisations of the fermion fields. We emphasise again that the field covariant formalism places no restriction on the space of allowed theories and therefore applies equally well to both Supersymmetric and non-Supersymmetric theories.

Let us review some of the key properties of supermanifolds that will be required to extend the field space covariant formalism to fermionic theories. This will provide a very basic introduction only. For further details on the mathematical structure of supermanifolds we refer the reader to [20, 21].

In order to set notation we shall consider a supermanifold with m commuting and n anticommuting coordinates. We shall denote the coordinates x^α, using Greek letters for the indices. When we need to refer to commuting and anticommuting coordinates separately we shall denote the former x^A with indices from the beginning of the Latin alphabet and the latter x^I with indices from the middle of the Latin alphabet. Thus, we have $\alpha \in \{1, \ldots, n+m\}$, while $A \in \{1, \ldots, n\}$ and $I \in \{1, \ldots, m\}$.

We shall also make use of the following convention, commonly found in the literature on supermanifolds. The exponent of a factor of -1 is not meant to be taken literally. Instead it should be taken as a label that gives 0 or 1 depending on the $\mathbb{Z}_2$ grading of the quantity it refers to. Thus, $(-1)^X$ is $+1$ when X is a commuting object and -1 when X is an anticommuting object. Similarly, $(-1)^\alpha$ is $+1$ when the index α refers to a commuting coordinate and -1 when α refers to an anticommuting coordinate. We note that an index in an exponent of -1 does not imply summation as would be expected in Einstein summation notation.

Many of the features of Riemannian manifolds discussed in Sect. 2.1 readily generalise to supermanifolds. However, the introduction of anticommuting degrees of freedom requires us to define things more carefully. In particular, when differentiating with respect to an anticommuting coordinate we must specify whether we are differentiating from the left or from the right. As we have already seen the left and right derivatives differ in general. However, it can be shown that they are related by

$$\overrightarrow{\partial}_\alpha X = (-1)^{\alpha(X+1)}\, X\,\overleftarrow{\partial}_\alpha. \tag{4.11}$$

Here, $\overrightarrow{\partial}_\alpha \equiv \overrightarrow{\partial}/\partial x^\alpha$ and $\overleftarrow{\partial}_\alpha \equiv \overleftarrow{\partial}/\partial x^\alpha$.

The distinction between left and right derivatives means that diffeomorphisms of the supermanifold

$$x^\alpha \to \widetilde{x}^\alpha = \widetilde{x}^\alpha(\boldsymbol{x}), \tag{4.12}$$

come with two distinct Jacobians

$${}_\alpha J^\beta = \frac{\overrightarrow{\partial}}{\partial x^\alpha}\widetilde{x}^\beta, \qquad\qquad {}^\beta J^{\mathsf{sT}}_\alpha = \widetilde{x}^\beta \frac{\overleftarrow{\partial}}{\partial x^\alpha}, \tag{4.13}$$

which we call the left and right Jacobian, respectively. Here, sT stands for the *supertranspose*, which we shall define in due course.

When working with tensors on the supermanifold we must, therefore, specify which of these Jacobians we should use in the transformation rule (2.1). We do this by placing the appropriate index on the right or left for tensors that transform with the left and right Jacobian, respectively. Thus, in general, a tensor will transform as

$$\begin{aligned}
{}^{\alpha_1\cdots\alpha_a}T \to {}^{\alpha_1\cdots\alpha_a}\widetilde{T} &= \left[(-1)^{\sum_{i\neq j}\alpha_i\alpha_j}\right] {}^{\alpha_a}J^{\mathsf{sT}}_{\gamma_a}\cdots{}^{\alpha_1}J^{\mathsf{sT}}_{\gamma_1}\,{}^{\gamma_1\cdots\gamma_a}T,\\
T^{\alpha_1\cdots\alpha_a} \to \widetilde{T}^{\alpha_1\cdots\alpha_a} &= \left[(-1)^{\sum_{i\neq j}\alpha_i\alpha_j}\right] T^{\gamma_1\cdots\gamma_a}\,{}_{\gamma_a}J^{\alpha_a}\cdots{}_{\gamma_a}J^{\alpha_1},\\
{}_{\alpha_1\cdots\alpha_a}T \to {}_{\alpha_1\cdots\alpha_a}\widetilde{T} &= \left[(-1)^{\sum_{i\neq j}\alpha_i\alpha_j}\right] {}_{\alpha_a}\!\left(J^{-1}\right)^{\gamma_a}\cdots{}_{\alpha_1}\!\left(J^{-1}\right)^{\gamma_1}\,{}_{\gamma_1\cdots\gamma_a}T,\\
T_{\alpha_1\cdots\alpha_a} \to \widetilde{T}_{\alpha_1\cdots\alpha_a} &= \left[(-1)^{\sum_{i\neq j}\alpha_i\alpha_j}\right] T_{\gamma_1\cdots\gamma_a}\,{}^{\gamma_a}\!\left(J^{-1}\right)^{\mathsf{sT}}_{\alpha_a}\cdots{}^{\gamma_a}\!\left(J^{-1}\right)^{\mathsf{sT}}_{\alpha_1}.
\end{aligned} \tag{4.14}$$

Note that we have separated the four types of indices for clarity but, in general, a tensor will have a mixture of these indices.

In addition we note that only indices that are immediately adjacent to one another can be contracted straightforwardly to produce a new tensor. To contract any non-adjacent indices we must introduce factors of -1 to account for the anticommutativity of the coordinates as we have in (4.14).

In (4.13) we introduced the supertranspose sT. The supertranspose generalises the notion of the transpose of a rank-2 tensor on a manifold (also known as a matrix). When we generalise to a rank-2 tensor on a supermanifold (also known as a supermatrix) we must introduce factors of -1 to take care of the commutation properties of the coordinates. These factors differ depending on the placement of the indices. In particular we have

$$\begin{aligned} {}_{\alpha}M^{\beta\,\mathsf{sT}} &= (-1)^{\beta(\alpha+1)}\;{}^{\beta}M_{\alpha}, \\ {}^{\alpha}M_{\beta}^{\mathsf{sT}} &= (-1)^{\beta(\alpha+1)}\;{}_{\beta}M^{\alpha}, \\ {}_{\alpha}M_{\beta}^{\mathsf{sT}} &= (-1)^{\alpha+\beta+\alpha\beta}\;{}_{\beta}M_{\alpha}, \\ {}^{\alpha}M^{\beta\,\mathsf{sT}} &= (-1)^{\alpha\beta}\;{}^{\beta}M^{\alpha}. \end{aligned} \tag{4.15}$$

The supertranspose satisfies the same identities one would expect of the regular transpose, namely

$$(M^{\mathsf{sT}})^{\mathsf{sT}} = M, \qquad (M^{-1})^{\mathsf{sT}} = (M^{\mathsf{sT}})^{-1}, \qquad (MN)^{\mathsf{sT}} = N^{\mathsf{sT}}M^{\mathsf{sT}}. \tag{4.16}$$

Note that, in the presence of anticommuting coordinates, the regular transpose *does not* satisfy all of these and, in particular, $(MN)^{\mathsf{T}} \neq N^{\mathsf{T}}M^{\mathsf{T}}$.

The definition of the supertranspose leads us to define supersymmetric and anti-supersymmetric supermatrices that satisfy

$$S^{\mathsf{sT}} = S \qquad \text{(supersymmetric)}, \tag{4.17}$$

$$A^{\mathsf{sT}} = -A \qquad \text{(antisupersymmetric)}, \tag{4.18}$$

respectively. We emphasise that supersymmetric in this context has nothing to do with the theory of Supersymmetry, but is merely the generalisation of the concept of a symmetric matrix in the presence of anticommuting coordinates. In order to distinguish between the two concepts we shall always refer to the physical theory as "Supersymmetry" with a capital S or by the acronym SUSY and use the lowercase "supersymmetric" when referring to the property (4.17).

We now wish to make our supermanifold Riemannian by equipping it with a metric ${}_{\alpha}g_{\beta}$. In doing so we can define a line element

$$ds^2 = dx^{\alpha}\;{}_{\alpha}g_{\beta}\;dx^{\beta}. \tag{4.19}$$

We require that the line element be a commuting number and, thus, the diagonal elements ${}_{A}g_{B}$ and ${}_{I}g_{J}$ are commuting numbers while the off-diagonal elements ${}_{A}g_{J}$ and ${}_{I}g_{B}$ are anticommuting numbers. Because of this, we see that ${}_{\alpha}g_{\beta}$ must be a supersymmetric supermatrix since any antisupersymmetric part would vanish in (4.19).

With the supermanifold equipped with a metric we can construct the Christoffel symbols

$${}^{\alpha}\Gamma_{\beta\gamma} = \frac{1}{2}{}^{\alpha}g^{\delta}\left[{}_{\delta}g_{\beta}\overleftarrow{\partial}_{\gamma} + (-1)^{\beta\gamma}\;{}_{\delta}g_{\gamma}\overleftarrow{\partial}_{\beta} - (-1)^{\beta}\overrightarrow{\partial}_{\delta}\;{}_{\beta}g_{\gamma}\right]. \tag{4.20}$$

This leads to the following definition of the left covariant derivative

$$\begin{aligned}
\overrightarrow{\nabla}_\beta T^{\alpha_1\cdots\alpha_a} &= \overrightarrow{\partial}_\beta T^{\alpha_1\cdots\alpha_a} + \sum_i \left[(-1)^{\beta(\alpha_i+1)+(\alpha_i+\gamma)\sum_{j<i}\alpha_j}\right] {}^{\alpha_i}\Gamma_{\beta\gamma} T^{\alpha_1\cdots\gamma\cdots\alpha_a}, \\
\overrightarrow{\nabla}_\beta T_{\alpha_1\cdots\alpha_a} &= \overrightarrow{\partial}_\beta T_{\alpha_1\cdots\alpha_a} - \sum_i \left[(-1)^{(\alpha_i+\beta)(\gamma+1)+(\alpha_i+\gamma)\sum_{j<i}\alpha_j}\right] {}^{\gamma}\Gamma_{\beta\alpha_i} T_{\alpha_1\cdots\gamma\cdots\alpha_a}, \\
\overrightarrow{\nabla}_\beta {}^{\alpha_1\cdots\alpha_a}T &= \overrightarrow{\partial}_\beta {}^{\alpha_1\cdots\alpha_a}T + \sum_i \left[(-1)^{\beta(\alpha_i+1)+(\alpha_i+\gamma)\sum_{j<i}\alpha_j}\right] {}^{\alpha_i}\Gamma_{\beta\gamma} {}^{\alpha_1\cdots\gamma\cdots\alpha_a}T, \\
\overrightarrow{\nabla}_\beta {}_{\alpha_1\cdots\alpha_a}T &= \overrightarrow{\partial}_\beta {}_{\alpha_1\cdots\alpha_a}T - \sum_i \left[(-1)^{\beta(\gamma+1)+\alpha_i+(\alpha_i+\gamma)\sum_{j<i}\alpha_j}\right] {}^{\gamma}\Gamma_{\beta\alpha_i} {}_{\alpha_1\cdots\gamma\cdots\alpha_a}T.
\end{aligned} \tag{4.21}$$

We can also define the right covariant derivative

$$\begin{aligned}
T^{\alpha_1\cdots\alpha_a}\overleftarrow{\nabla}_\beta &= T^{\alpha_1\cdots\alpha_a}\overleftarrow{\partial}_\beta + \sum_i \left[(-1)^{\gamma\alpha_i+(\alpha_i+\gamma)\sum_{j<i}\alpha_j}\right] T^{\alpha_1\cdots\gamma\cdots\alpha_a}\, {}^{\alpha_i}\Gamma_{\gamma\beta}, \\
T_{\alpha_1\cdots\alpha_a}\overleftarrow{\nabla}_\beta &= T_{\alpha_1\cdots\alpha_a}\overleftarrow{\partial}_\beta - \sum_i \left[(-1)^{(\alpha_i+\gamma)\sum_{j<i}\alpha_j}\right] T_{\alpha_1\cdots\gamma\cdots\alpha_a}\, {}^{\gamma}\Gamma_{\alpha_i\beta}, \\
{}^{\alpha_1\cdots\alpha_a}T\overleftarrow{\nabla}_\beta &= {}^{\alpha_1\cdots\alpha_a}T\overleftarrow{\partial}_\beta + \sum_i \left[(-1)^{\gamma\alpha_i+(\alpha_i+\gamma)\sum_{j<i}\alpha_j}\right] {}^{\alpha_1\cdots\gamma\cdots\alpha_a}T\, {}^{\alpha_i}\Gamma_{\gamma\beta}, \\
{}_{\alpha_1\cdots\alpha_a}T\overleftarrow{\nabla}_\beta &= {}_{\alpha_1\cdots\alpha_a}T\overleftarrow{\partial}_\beta - \sum_i \left[(-1)^{\alpha_i(\gamma+1)+\gamma+(\alpha_i+\gamma)\sum_{j<i}\alpha_j}\right] {}_{\alpha_1\cdots\gamma\cdots\alpha_a}T\, {}^{\gamma}\Gamma_{\alpha_i\beta}.
\end{aligned} \tag{4.22}$$

Again we have separated the four types of indices for clarity, but the definition generalises straightforwardly to tensors with mixed indices.

We can also define the curvature invariants in the usual way. However, we must be careful to take account of the ordering of anticommuting quantities and to insert the correct factors of -1. We find that the generalisations of (2.16), (2.19) and (2.20) are, respectively [20],

$$\begin{aligned}
{}^{\alpha}R_{\beta\gamma\delta} =& - {}^{\alpha}\Gamma_{\beta\gamma}\overleftarrow{\partial}_\delta + (-1)^{\gamma\delta}\, {}^{\alpha}\Gamma_{\beta\delta}\overleftarrow{\partial}_\gamma \\
&+ (-1)^{\gamma(\beta+\epsilon)}\, {}^{\alpha}\Gamma_{\epsilon\gamma}\, {}^{\epsilon}\Gamma_{\beta\delta} - (-1)^{\delta(\epsilon+\beta+\gamma)}\, {}^{\alpha}\Gamma_{\epsilon\delta}\, {}^{\epsilon}\Gamma_{\beta\gamma},
\end{aligned} \tag{4.23}$$

$$R_{\alpha\beta} = (-1)^{\gamma(\alpha+1)}\, {}^{\gamma}R_{\alpha\gamma\beta}, \tag{4.24}$$

$$R = R_{\alpha\beta}\, {}^{\beta}g^{\alpha}. \tag{4.25}$$

Here, ${}^{\alpha}g^{\beta}$ is the inverse metric satisfying

$${}^{\alpha}g^{\gamma}\, {}_{\gamma}g_{\beta} = {}^{\alpha}\delta_{\beta}. \tag{4.26}$$

Finally, we wish to define integrals on the supermanifold in a diffeomorphism invariant way. This can be achieved using the *superdeterminant* of the metric, also known as the Berezinian [11]. For a supermatrix of the form

$$ {}_{\alpha}M_{\beta} = \begin{pmatrix} {}_{A}A_{B} & {}_{A}C_{J} \\ {}_{I}D_{B} & {}_{I}B_{J} \end{pmatrix} \quad (4.27) $$

the superdeterminant is defined to be

$$ \operatorname{sdet} M = \frac{\det(A - CB^{-1}D)}{\det B}. \quad (4.28) $$

With this definition the Berezinian integral measure

$$ dV = \sqrt{\operatorname{sdet}(g)}\, d^{n+m}x \quad (4.29) $$

is invariant under (4.12).

4.2 The Fermionic Field Space

We can now construct the field space for a theory with N scalar fields ϕ^A (collectively ϕ) and M Dirac fermions ψ^X. We recall that, in 4 dimensions, each Dirac fermion carries 4 complex degrees of freedom [7], and therefore the total number of real degrees of freedom in this theory is $N + 8M$. The appropriate field space is, thus, a supermanifold with N commuting and $8M$ anticommuting coordinates.

We can describe the field space with the following set of coordinates

$$ \Phi^{\alpha} = \left(\phi^A,\ \psi^1_a,\ \overline{\psi}^1_{\dot{a}},\ \psi^2_a,\ \overline{\psi}^2_{\dot{a}},\ \ldots\right), \quad (4.30) $$

where the subscripts a and $\dot{a}$ refer to the spinor components of the Dirac fermions and $\overline{\psi}^X \equiv \psi^{X*}\gamma^0$. We note that the index α (as well as all other indices denoted with Greek letters) runs over all N scalar fields as well as all 8 degrees of freedom for each fermion field, including the spinor components. Thus, we have $\alpha \in \{1, \ldots, N + 8M\}$. As before, when we wish to refer to the commuting and anticommuting coordinates separately we shall use capital letters from the beginning of the Latin alphabet for the former and capital letters from the middle of the Latin alphabet for the latter. We, therefore, have $A \in \{1, \ldots, N\}$ and $I \in \{1, \ldots, 8M\}$. In addition, capital letters from the end of the Latin alphabet will denote the fermion flavour with $X \in \{1, \ldots, M\}$ and lower case Latin letters will be used to denote spinor indices with $a \in \{1, 2, 3, 4\}$.

Diffeomorphisms of the field space take the form

$$ \Phi^{\alpha} \to \widetilde{\Phi}^{\alpha} = \widetilde{\Phi}^{\alpha}(\Phi). \quad (4.31) $$

We see that such diffeomorphisms cover a wide range of field redefinitions. In fact any redefinition of the scalar and fermion fields that does not mix derivative and non-derivative terms[1] will be covered by (4.31). An example of such a transformation is

$$\phi^A \to \widetilde{\phi}^A = \widetilde{\phi}^A(\phi), \qquad \psi^X \to \widetilde{\psi}^X = (K(\phi)^{-1})^X_Y \, \psi^Y, \tag{4.32}$$

which is the most common type of field redefinition for such theories. Thus, as before, we can impose manifest reparametrisation invariance for fermionic theories by constructing our theory out of tensors on the field space supermanifold.

4.3 Field Space Covariant Lagrangians for Fermionic Theories

Let us now write down a Lagrangian for our scalar fermion theory and investigate its field space properties. Allowing up to two derivatives for the scalar fields, and one derivative for the fermionic fields, the most general Lagrangian for this theory is

$$\begin{aligned}\mathcal{L} =& \frac{1}{2} g^{\mu\nu} k_{AB}(\boldsymbol{\Phi}) \partial_\mu \phi^A \partial_\nu \phi^B - \frac{1}{2} h_{AXY}(\boldsymbol{\Phi}) \overline{\psi}^X \gamma^\mu \psi^Y \partial_\mu \phi^A \\ &+ \frac{i}{2} g_{XY}(\boldsymbol{\Phi}) \left(\overline{\psi}^X \gamma^\mu \partial_\mu \psi^Y - \partial_\mu \overline{\psi}^X \gamma^\mu \psi^Y \right) \\ &+ \frac{i}{2} j_{WXYZ}(\boldsymbol{\Phi}) \overline{\psi}^W \gamma^\mu \psi^X \left(\overline{\psi}^Y \partial_\mu \psi^Z - \partial_\mu \overline{\psi}^Y \psi^Z \right) \\ &- Y_{XY}(\boldsymbol{\Phi}) \overline{\psi}^X \psi^Y - V(\phi).\end{aligned} \tag{4.33}$$

Here, k_{AB}, h_{AXY}, g_{XY}, j_{WXYZ} and Y_{XY} are arbitrary functions that can, in general, depend on both the scalar fields and bilinears of the fermion fields.

We note that X, Y, Z, etc. denote the fermion flavour and are not field space indices. Hence, the model functions defined by (4.33) are not field space covariant objects. Let us, therefore, rewrite this Lagrangian, absorbing these non-covariant terms into field space tensors, so that the expression becomes reparametrisation invariant. This leads us to define

[1] This means that the results of this chapter do not extend to SUSY transformations [23] since these include a dependence on the field derivatives. We shall discuss such transformations in Chap. 8.

$$
\begin{aligned}
{}_{\alpha}k_{\beta} &= \begin{pmatrix} k_{AB} & \mathbf{0}_{N\times 8M} \\ \mathbf{0}_{8M\times N} & \mathbf{0}_{8M\times 8M} \end{pmatrix}, \\
\zeta^{\mu}_{\alpha} &= \begin{pmatrix} ih_{AXY}\,\overline{\psi}^{X}\gamma^{\mu}\psi^{Y} \\ g_{YX}\,\overline{\psi}^{Y}\gamma^{\mu} + j_{WYXZ}\overline{\psi}^{W}\gamma^{\mu}\psi^{Y}\overline{\psi}^{Z} \\ g_{XY}\,\gamma^{\mu}\psi^{Y} + j_{WYZX}\overline{\psi}^{W}\gamma^{\mu}\psi^{Y}\psi^{Z} \end{pmatrix}, \\
U &= Y_{XY}\overline{\psi}^{X}\psi^{Y} + V,
\end{aligned}
\tag{4.34}
$$

and, thus, write the Lagrangian as

$$
\mathcal{L} = \frac{1}{2}g^{\mu\nu}\partial_{\mu}\Phi^{\alpha}\ {}_{\alpha}k_{\beta}(\mathbf{\Phi})\ \partial_{\nu}\Phi^{\beta} + \frac{i}{2}\zeta^{\mu}_{\alpha}(\mathbf{\Phi})\ \partial_{\mu}\Phi^{\alpha} - U(\mathbf{\Phi}). \tag{4.35}
$$

This Lagrangian contains three model functions, ${}_{\alpha}k_{\beta}$, ζ^{μ}_{α} and U, which are all field space tensors and thus (4.35) is a reparametrisation invariant expression.

As well as the expressions given in (4.34), we can also define these three tensors directly from the Lagrangian with the following equations:

$$
{}_{\alpha}k_{\beta} = \frac{g_{\mu\nu}}{4}\frac{\overrightarrow{\partial}}{\partial(\partial_{\mu}\Phi^{\alpha})}\ \mathcal{L}\ \frac{\overleftarrow{\partial}}{\partial(\partial_{\nu}\Phi^{\beta})}, \tag{4.36}
$$

$$
\zeta^{\mu}_{\alpha} = \frac{2}{i}\left(\mathcal{L} - \frac{1}{2}g^{\mu\nu}\partial_{\mu}\Phi^{\alpha}\ {}_{\alpha}k_{\beta}\ \partial_{\nu}\Phi^{\beta}\right)\frac{\overleftarrow{\partial}}{\partial(\partial_{\mu}\Phi^{\alpha})}, \tag{4.37}
$$

$$
U = \frac{1}{2}g^{\mu\nu}\partial_{\mu}\Phi^{\alpha}\ {}_{\alpha}k_{\beta}(\mathbf{\Phi})\ \partial_{\nu}\Phi^{\beta} + \frac{i}{2}\zeta^{\mu}_{\alpha}(\mathbf{\Phi})\ \partial_{\mu}\Phi^{\alpha} - \mathcal{L}. \tag{4.38}
$$

Such a constructive definition is important in ensuring our formalism is uniquely defined.

In order to highlight the ubiquity of (4.35) let us end this section by considering a free scalar fermion theory with Lagrangian

$$
\begin{aligned}
\mathcal{L} = &\sum_{A\in\text{scalars}}\left[\frac{1}{2}g^{\mu\nu}\partial_{\mu}\phi^{A}\partial_{\nu}\phi^{A} - \frac{1}{2}m_{A}^{2}(\phi^{A})^{2}\right] \\
&+ \sum_{X\in\text{fermions}}\left[\frac{i}{2}\left(\overline{\psi}^{X}\gamma^{\mu}\partial_{\mu}\psi^{X} - \partial_{\mu}\overline{\psi}^{X}\gamma^{\mu}\psi^{X}\right) - m_{X}\overline{\psi}^{X}\psi^{X}\right].
\end{aligned}
\tag{4.39}
$$

We see that such a Lagrangian can be fully described by (4.35) if we choose the model functions to be

$$
\begin{aligned}
{}_{\alpha}k_{\beta} &= \begin{pmatrix} \delta_{AB} & \mathbf{0}_{N\times 8M} \\ \mathbf{0}_{8M\times N} & \mathbf{0}_{8M\times 8M} \end{pmatrix}, \\
\zeta^{\mu}_{\alpha} &= \left(\mathbf{0}_{N},\ \overline{\psi}^{1}\gamma^{\mu},\ \gamma^{\mu}\psi^{1},\ \overline{\psi}^{2}\gamma^{\mu},\ \gamma^{\mu}\psi^{2},\ \ldots\right), \\
U &= \sum_{A\in\text{scalars}} \frac{1}{2}m_A^2(\phi^A)^2 + \sum_{X\in\text{fermions}} m_X\overline{\psi}^{X}\psi^{X}.
\end{aligned}
\tag{4.40}
$$

4.4 Tensors on the Field Space

Let us consider the three tensors in (4.34) one at a time. The quantity ${}_{\alpha}k_{\beta}$ is a rank $\binom{0}{2}$ field space tensor. In a pure scalar theory, this tensor acts as the field space metric. However, in the presence of fermions this is no longer the case. Fermions enter the Lagrangian with only a single derivative and thus ${}_{\alpha}k_{\beta}$ has no support in the fermionic sector, i.e.

$$
{}_{\alpha}k_{I} = {}_{I}k_{\alpha} = 0. \tag{4.41}
$$

Hence, ${}_{\alpha}k_{\beta}$ is singular and so cannot be used as the field space metric.

Next we consider $U(\mathbf{\Phi})$. This term contains the scalar potential $V(\phi)$ as well as the fermion mass terms and any momentum independent interactions between the scalars and fermions, e.g. Yukawa interactions [30]. As before, the potential term will not play a role in the construction of the field space.

Finally, we consider ζ^{μ}_{α}, which is both a field space covector and a spacetime vector. This model function has no analogue in pure scalar theories since such theories do not contain kinetic terms with a single derivative. The reason for this is the following; since ζ^{μ}_{α} cannot depend on derivatives and there are no vector fields in our theory, the spacetime index μ of ζ^{μ}_{α} must be inherited from a γ^{μ} matrix. It, therefore, cannot appear in a theory without fermions since there would be nothing to contract with the spinor indices of γ^{μ}.

Note that the dependence on γ^{μ} *does not* imply that ζ^{μ}_{α} only has support in the fermionic sector. Indeed, we see from (4.34) that $\zeta^{\mu}_{A} \neq 0$ as long as h_{AXY} has some non-zero components. This is a very generic condition as, even if $h_{AXY} = 0$ in some parametrisation, it may not be zero after a field redefinition (4.31). For example, if we take the canonical theory (4.39) and perform the transformation (4.32), we will end up with

$$
h_{AXY} = i\sum_{Z}\left(K^{Z}_{Y}\partial_{A}K^{*Z}_{X} - K^{*Z}_{X}\partial_{A}K^{Z}_{Y}\right) \neq 0. \tag{4.42}
$$

Since we know that the spacetime transformation properties of ζ^{μ}_{α} always stem from a γ^{μ} matrix, it seems that we could define a pure field space vector if we could find a way to surgically remove these matrices. Whilst the simplest way to achieve this would be to contract ζ^{μ}_{α} with a γ^{μ} matrix and rely on the Clifford algebra, this is not possible in general because ζ^{μ}_{α} does not have the right spinor structure.

We shall, therefore, instead remove the γ^μ matrices by defining the notion of differentiation with respect to γ^μ. Such a process can be defined in a rigorous way be recalling that any matrix in spinor space can be written uniquely in terms of 16 Lorentz-covariant bilinears as [31]

$$M_{a\dot{a}} = \sum_{i=1}^{16} a^{(i)}\,\Gamma^{(i)}_{a\dot{a}}, \qquad \Gamma^{(i)} \in \{I_4, \gamma^5, \gamma^\mu, \gamma^\mu\gamma^5, \sigma^{\mu\nu}\}, \tag{4.43}$$

where $\Gamma^{(i)}$ are the basis matrices, $a^{(i)}$ is a coefficient, and $\sigma^{\mu\nu} \equiv i/2\,[\gamma^\mu, \gamma^\nu]$. Thus, we can define partial differentiation with respect to one of these matrices as

$$\frac{\delta F}{\delta \Gamma^{(i)}} \equiv \lim_{\epsilon^{(i)}\to 0} \frac{F[\Gamma^{(i)} \to \Gamma^{(i)} + \epsilon^{(i)} I_4] - F[\Gamma^{(i)}]}{\epsilon^{(i)}}, \tag{4.44}$$

such that, when applied to (4.43) we have

$$\frac{\delta M_{a\dot{a}}}{\delta \Gamma^{(i)}} = a^{(i)}\delta_{a\dot{a}}. \tag{4.45}$$

We can now use (4.44) to define[2]

$$\zeta_\alpha \equiv \frac{1}{4}\frac{\delta \zeta^\mu_\alpha}{\delta \gamma^\mu}, \tag{4.46}$$

which is equivalent to replacing all γ^μ matrices with the identity. For example, for the canonical theory (4.39) this definition gives us

$$\zeta_\alpha = \Big(\mathbf{0}_N,\ \overline{\psi}^1,\ \psi^1,\ \overline{\psi}^2,\ \psi^2,\ \ldots\Big). \tag{4.47}$$

The quantity ζ_α defined in this way is a pure field-space covector and transforms as

$$\zeta_\alpha \to \widetilde{\zeta}_\alpha = \zeta_\beta\ {}^\beta\big(J^{-1}\big)^{\mathsf{sT}}_\alpha \tag{4.48}$$

under a field redefinition (4.31), where

$${}_\alpha J^\beta = \frac{\overrightarrow{\partial}}{\partial \Phi^\alpha}\widetilde{\Phi}^\beta \tag{4.49}$$

is the left Jacobian of the transformation.

We can use the vector ζ_α to define a rank $\binom{0}{2}$ tensor on the field space supermanifold

[2] The factor of 1/4 is included in order to compensate for the factor of 4 arising from the contraction of spacetime indices.

$$ {}_{\alpha}\lambda_{\beta} = \frac{\overrightarrow{\partial}}{\partial \Phi^{\alpha}} \zeta_{\beta} - (-1)^{\alpha+\beta+\alpha\beta} \frac{\overrightarrow{\partial}}{\partial \Phi^{\beta}} \zeta_{\alpha}. \tag{4.50} $$

Despite the appearance of ordinary partial derivatives in this expression, ${}_{\alpha}\lambda_{\beta}$ is a genuine tensor and transforms as

$$ {}_{\alpha}\lambda_{\beta} \to {}_{\alpha}\widetilde{\lambda}_{\beta} = {}_{\alpha}\left(J^{-1}\right)^{\gamma}\ {}_{\gamma}\lambda_{\delta}\ {}^{\delta}\left(J^{-1}\right)^{\mathsf{sT}}_{\beta}. \tag{4.51} $$

under a field redefinition (4.31).

The reason for this is analogous to the reason that the field strength tensor $F_{\mu\nu}$ transforms as a spacetime tensor in QED [32–34]. If we were to replace the partial derivatives in (4.50) with covariant derivatives, any connection terms would cancel. Notice that, similarly to the field strength tensor, the tensor ${}_{\alpha}\lambda_{\beta}$ is antisupersymmetric with $\lambda = -\lambda^{\mathsf{sT}}$.

Let us highlight another important feature of ${}_{\alpha}\lambda_{\beta}$. Consider adding to the Lagrangian (4.35) a total derivative of the form

$$ \mathcal{L}_{\mathrm{TD}} = \frac{i}{2}\partial_{\mu}\left(t_{XY}(\phi)\overline{\psi}^{X}\gamma^{\mu}\psi^{Y}\right), \tag{4.52} $$

where $t_{XY} = t_{YX}$ is a real symmetric matrix. Because the definition (4.50) is linear we see that ${}_{\alpha}\lambda_{\beta}$ for the full Lagrangian is

$$ {}_{\alpha}\lambda_{\beta}\left[\mathcal{L} + \mathcal{L}_{\mathrm{TD}}\right] = {}_{\alpha}\lambda_{\beta}\left[\mathcal{L}\right] + {}_{\alpha}\lambda_{\beta}\left[\mathcal{L}_{\mathrm{TD}}\right]. \tag{4.53} $$

But, for $\mathcal{L}_{\mathrm{TD}}$ we have

$$ \zeta_{\alpha}\left[\mathcal{L}_{\mathrm{TD}}\right] = \left(\partial_{A} t_{XY}\overline{\psi}^{X}\psi^{Y},\ \ t_{YX}\overline{\psi}^{Y},\ \ -t_{XY}\psi^{Y}\right) \tag{4.54} $$

and thus

$$ {}_{\alpha}\lambda_{\beta}\left[\mathcal{L}_{\mathrm{TD}}\right] = 0. \tag{4.55} $$

Therefore, ${}_{\alpha}\lambda_{\beta}$ is unchanged if total derivative terms are added to the Lagrangian. This is important since such terms do not affect the dynamics of the theory.

The tensor ${}_{\alpha}\lambda_{\beta}$ is singular for most common field theories. For example, in the case of the canonical theory (4.39) we have

$$ {}_{\alpha}\lambda_{\beta} = \begin{pmatrix} \mathbf{0}_{N\times N} & 0 & 0 & 0 & 0 & \cdots \\ 0 & 0 & 1_4 & 0 & 0 & \cdots \\ 0 & 1_4 & 0 & 0 & 0 & \cdots \\ 0 & 0 & 0 & 0 & 1_4 & \cdots \\ 0 & 0 & 0 & 1_4 & 0 & \cdots \\ \vdots & \vdots & \vdots & \vdots & \vdots & \ddots \end{pmatrix}. \tag{4.56} $$

This limits the usefulness of this tensor since it has no inverse and its superdeterminant is zero.

Thus, it will be useful to define

$$ {}_{\alpha}\Lambda_{\beta} = {}_{\alpha}k_{\beta} + {}_{\alpha}\lambda_{\beta}. \tag{4.57} $$

For the canonical theory (4.39) this definition gives

$$ {}_{\alpha}\Lambda_{\beta} = {}_{\alpha}N_{\beta} \equiv \begin{pmatrix} 1_N & 0 & 0 & 0 & 0 & \cdots \\ 0 & 0 & 1_4 & 0 & 0 & \cdots \\ 0 & 1_4 & 0 & 0 & 0 & \cdots \\ 0 & 0 & 0 & 0 & 1_4 & \cdots \\ 0 & 0 & 0 & 1_4 & 0 & \cdots \\ \vdots & \vdots & \vdots & \vdots & \vdots & \ddots \end{pmatrix}. \tag{4.58} $$

We see that ${}_{\alpha}\Lambda_{\beta}$ is now non-singular and, therefore, has an inverse and a non-zero superdeterminant.

4.5 The Field Space Metric

We now have a field space supermanifold equipped with some tensors. However, in order to use the full power of field space covariance and, in particular, the VDW effective action, we must make the field space Riemannian by equipping it with a metric. We wish to define the metric in such a way that it satisfies the following properties:

1. The metric should be determined solely and uniquely from the action.
2. The metric should transform as a rank $\binom{0}{2}$ field-space tensor.
3. The metric should be supersymmetric.
4. The metric should not be singular, unless there are non-dynamical degrees of freedom.
5. The metric should depend on the fields only and not on their derivatives.
6. The metric for a theory with canonically normalised fields should be given by

$$ {}_{a}H_{b} \equiv \begin{pmatrix} 1_N & 0 & 0 & 0 & 0 & \cdots \\ 0 & 0 & 1_4 & 0 & 0 & \cdots \\ 0 & -1_4 & 0 & 0 & 0 & \cdots \\ 0 & 0 & 0 & 0 & 1_4 & \cdots \\ 0 & 0 & 0 & -1_4 & 0 & \cdots \\ \vdots & \vdots & \vdots & \vdots & \vdots & \ddots \end{pmatrix}, \tag{4.59} $$

which is the analogue of the Euclidean metric for a supermanifold [20].[3]

In the previous section we defined a tensor ${}_\alpha\Lambda_\beta$, which satisfies some of these properties. It is uniquely determined from the action, it transforms as a rank two tensor, it is non-singular and it does not depend on derivatives of the field. However, $\Lambda \neq \Lambda^{\mathsf{sT}}$ and so this tensor does not satisfy property 3. As we have seen before in Sect. 4.1, only the supersymmetric part of the metric contributes to the line element. We might be tempted to use supersymmetric part of ${}_\alpha\Lambda_\beta$, but this is ${}_\alpha k_\beta$, which, as we have argued before, is singular and so violates property 4. We, therefore, need to define the metric another way.

Instead of focusing on the metric directly, let us consider the field space vielbeins ${}_\alpha e^a$. Given a set of vielbeins, we can write the desired metric as

$$ {}_\alpha G_\beta = {}_\alpha e^a \; {}_a H_b \; {}^b e^{\mathsf{sT}}_\beta . \tag{4.60} $$

As we have seen in Sect. 2.1.6, defining the vielbeins is an equivalent way to define the metric.

However, the vielbeins also have another interpretation. They are the Jacobian of the transformation to the local inertial frame at a given point on the field space. Thanks to property 6 we know that a flat field space corresponds to canonical kinetic terms and so in the local inertial frame the Lagrangian must be locally equivalent to (4.39) (up to a possible irrelevant change in the potential term). We, therefore, see that ${}_a\Lambda_b = {}_a N_b$ in the local inertial frame and thus

$$ {}_\alpha \Lambda_\beta = {}_\alpha e^a \; {}_a N_b \; {}^b e^{\mathsf{sT}}_\beta . \tag{4.61} $$

We can calculate ${}_\alpha\Lambda_\beta$ directly from the previous section and therefore the only unknowns in (4.61) are the vielbeins. We can, therefore, solve this equation to find the field space vielbeins and consequently calculate the field space metric.

We note, however, that (4.61) is not sufficient to uniquely define the vielbeins. Indeed, we can generate a new solution to (4.61) by multiplying any existing solution by a supermatrix ${}_a X^b$ that satisfies

$$ {}_a X^c \; {}_c N_d \; {}^d X^{\mathsf{sT}}_b = {}_a N_b . \tag{4.62} $$

The most general such supermatrix can be written as a product

$$ {}_a X^b = {}_a Y^c \; {}_c X^b_0 , \tag{4.63} $$

where ${}_c X^b_0$ is given by

---

[3] Note that we have assumed the commuting part of the metric has a purely positive signature, indicating no ghost fields in the scalar sector. If the theory were to include ghosts, we would have to replace the Euclidean metric 1_N in the commuting part of ${}_a H_b$ with the appropriate pseudo Euclidean metric η.

$$
{}_aX_0^b = \begin{pmatrix} O_N & 0 & 0 & 0 & 0 & \cdots \\ 0 & x_1 & 0 & 0 & 0 & \cdots \\ 0 & 0 & x_1^{-1} & 0 & 0 & \cdots \\ 0 & 0 & 0 & x_2 & 0 & \cdots \\ 0 & 0 & 0 & 0 & x_2^{-1} & \cdots \\ \vdots & \vdots & \vdots & \vdots & \vdots & \ddots \end{pmatrix}. \tag{4.64}
$$

Here, O_N is an orthogonal $N \times N$ matrix and x_X are a set of M arbitrary invertible 4×4 matrices. We note that ${}_aX_0^c\,{}_cH_d\,{}^dX^{sT}_{0b} = {}_aH_b$ regardless of the values of O_N and x_X and, therefore, these choices will not affect the metric (4.60) and can be chosen arbitrarily without violating property 1.

The supermatrix ${}_aY^b$ encodes the transformation $\psi \leftrightarrow \overline{\psi}$ for some set of fermion fields. It, therefore, takes on one of 2^M distinct possible values, which have the form

$$
{}_aY^b = \begin{pmatrix} 1_N & 0 & 0 & \cdots & 0 \\ 0 & y_1 & 0 & \cdots & 0 \\ 0 & 0 & y_2 & \cdots & 0 \\ \vdots & \vdots & \vdots & \ddots & \vdots \\ 0 & 0 & 0 & \cdots & y_M \end{pmatrix}, \tag{4.65}
$$

where each of the y_X is equal to either

$$
y_X = \begin{pmatrix} 1_4 & 0 \\ 0 & 1_4 \end{pmatrix} \quad \text{or} \quad \begin{pmatrix} 0 & 1_4 \\ 1_4 & 0 \end{pmatrix}. \tag{4.66}
$$

We notice that ${}_aN_b$ is invariant under such an exchange but ${}_aH_b$ is not, picking up extra minus signs in the $\psi\overline{\psi}$ and $\overline{\psi}\psi$ entries. Thus, (4.61) does not define one unique metric, but rather 2^M different ones.

However, only one of these metrics will have the correct signature. We see from (4.59), that all the entries on upper right diagonal of the metric are positive, whereas those on the lower left diagonal are negative. The locations of these minus signs constitute the signature of the metric of a supermanifold. These locations are a global property of the supermanifold, even for a non-trivial field space. The entries cannot take different signs in different parts of the supermanifold, since they would then have to pass through zero at which point the metric would become singular, contradicting property 4.

Only one choice of ${}_aY^b$ yields a metric with all positive entries in the upper right and all negative entries in the lower left. Any other choice would lead to a difference between the signature of the local metric and the global one. Such a difference is unacceptable for a proper supermanifold, just as it would be for a regular manifold. Therefore, the value of ${}_aY^b$ is entirely fixed and so the metric is unique after all, in agreement with property 1.

4.6 Unique Frame Invariant Effective Action for Fermions

With the Riemannian field space thus defined we can proceed to use the techniques of Vilkovisky and DeWitt to construct an effective action that is invariant under (4.31). We start be defining a connection on the field space

$$
{}^{\alpha}\Gamma_{\beta\gamma} = \frac{1}{2}{}^{\alpha}G^{\delta}\left[{}_{\delta}G_{\beta}\overleftarrow{\partial}_{\gamma} + (-1)^{\beta\gamma}\,{}_{\delta}G_{\gamma}\overleftarrow{\partial}_{\beta} - (-1)^{\beta}\overrightarrow{\partial}_{\delta}\,{}_{\beta}G_{\gamma}\right] \tag{4.67}
$$

as well as a field space covariant derivative

$$
X^{\alpha}\overleftarrow{\nabla}_{\beta} = X^{\alpha}\frac{\overleftarrow{\partial}}{\partial\Phi^{\beta}} + {}^{\alpha}\Gamma_{\beta\gamma}X^{\gamma}, \qquad X_{\alpha}\overleftarrow{\nabla}_{\beta} = X_{\alpha}\frac{\overleftarrow{\partial}}{\partial\Phi^{\beta}} - X_{\gamma}\,{}^{\gamma}\Gamma_{\alpha\beta}, \tag{4.68}
$$

with the generalisation to higher order tensors given by (4.22) and generalisation to left derivatives given by (4.21).

To continue, we must generalise the field space defined thus far to the infinite dimensional configuration space. Since we are considering gravity to be a background field and not a dynamical degree of freedom, we will not have to worry about the subtleties arising from the cosmological frame problem. We can, therefore, perform this generalisation straightforwardly by considering a set of coordinates

$$
\Phi^{\widehat{\alpha}} \equiv \Phi^{\alpha}(\boldsymbol{x}_{\alpha}), \tag{4.69}
$$

and taking the metric to be

$$
{}_{\widehat{\alpha}}G_{\widehat{\beta}} = {}_{\alpha}G_{\beta}\delta^{(4)}(\boldsymbol{x}_{\alpha} - \boldsymbol{x}_{\beta}). \tag{4.70}
$$

The configuration space connections are, therefore, given by

$$
\begin{aligned}
{}^{\widehat{\alpha}}\Gamma_{\widehat{\beta}\widehat{\gamma}} &\equiv \frac{1}{2}{}^{\widehat{\alpha}}G^{\widehat{\delta}}\left[{}_{\widehat{\delta}}G_{\widehat{\beta}}\frac{\overleftarrow{\delta}}{\delta\phi^{\widehat{\gamma}}} + (-1)^{\beta\gamma}\,{}_{\widehat{\delta}}G_{\widehat{\gamma}}\frac{\overleftarrow{\delta}}{\delta\phi^{\widehat{\beta}}} - (-1)^{\beta}\frac{\overrightarrow{\delta}}{\delta\phi^{\widehat{\delta}}}\,{}_{\widehat{\beta}}G_{\widehat{\gamma}}\right] \\
&= {}^{\alpha}\Gamma_{\beta\gamma}\delta^{(4)}(\boldsymbol{x}_{\alpha} - \boldsymbol{x}_{\beta})\delta^{(4)}(\boldsymbol{x}_{\alpha} - \boldsymbol{x}_{\gamma})
\end{aligned} \tag{4.71}
$$

and, hence, configuration space covariant derivatives are

$$
X^{\widehat{\alpha}}\overleftarrow{\nabla}_{\widehat{\beta}} = X^{\widehat{\alpha}}\frac{\overleftarrow{\delta}}{\delta\phi^{\widehat{\beta}}} + {}^{\widehat{\alpha}}\Gamma_{\widehat{\beta}\widehat{\gamma}}X^{\widehat{\gamma}}, \qquad X_{\widehat{\alpha}}\overleftarrow{\nabla}_{\widehat{\beta}} = X_{\widehat{\alpha}}\frac{\overleftarrow{\delta}}{\delta\phi^{\widehat{\beta}}} - X_{\widehat{\gamma}}\,{}^{\widehat{\gamma}}\Gamma_{\widehat{\beta}\widehat{\alpha}}. \tag{4.72}
$$

Again, generalisation to higher order tensors and the left derivative can be inferred from (4.21) and (4.22).

The configuration space formalism allows us to define a reparametrisation invariant path integral measure

$$[\mathcal{D}\mathcal{V}] = \sqrt{|\mathrm{sdet}\, G|}\;[\mathcal{D}\Phi]\,. \tag{4.73}$$

Notice that in the local inertial frame $|\mathrm{sdet}\, H| = |\mathrm{sdet}\, N|$. Hence, in any other frame we have $|\mathrm{sdet}\, G| = |\mathrm{sdet}\, \Lambda|$, since both quantities transform the same way under (4.31). We can, therefore, always replace G with Λ in (4.73) if this proves easier to calculate.

We can also define Vilkovisky's tangent vector [1, 2]

$$\sigma^{\widehat{\alpha}}[\varphi, \boldsymbol{\Phi}] = (\Phi^{\widehat{\alpha}} - \varphi^{\widehat{\alpha}}) + \frac{1}{2}\Gamma^{\widehat{\alpha}}_{\widehat{\beta}\widehat{\gamma}}[\varphi](\varphi^{\widehat{\gamma}} - \Phi^{\widehat{\gamma}})(\varphi^{\widehat{\beta}} - \Phi^{\widehat{\beta}}) + \cdots . \tag{4.74}$$

as well as Dewitt's modification [3]

$$\Sigma^{\widehat{\alpha}}[\varphi, \boldsymbol{\Phi}] = (C^{-1}[\varphi])^{\widehat{\alpha}}_{\widehat{\beta}}\, \sigma^{\widehat{\beta}}[\varphi, \boldsymbol{\Phi}], \tag{4.75}$$

where

$$C^{\widehat{\alpha}}_{\widehat{\beta}}[\varphi] = -\langle \sigma^{\widehat{\alpha}}[\varphi, \phi]\overleftarrow{\nabla}_{\widehat{\beta}}\rangle = \langle \delta^{\widehat{\alpha}}_{\widehat{\beta}} - (-1)^{\widehat{\beta}\widehat{\gamma}}\frac{1}{3}R^{\widehat{\alpha}}_{\widehat{\gamma}\widehat{\beta}\widehat{\delta}}[\varphi]\sigma^{\widehat{\delta}}[\varphi, \boldsymbol{\Phi}]\,\sigma^{\widehat{\gamma}}[\varphi, \boldsymbol{\Phi}] + \ldots\rangle. \tag{4.76}$$

As in the bosonic case, the choice (4.76) ensures that

$$\langle \Sigma(\varphi, \boldsymbol{\Phi})\rangle = 0 \tag{4.77}$$

and, thus, all tadpole diagrams will vanish as discovered in Sect. 2.5.2. In particular, we have

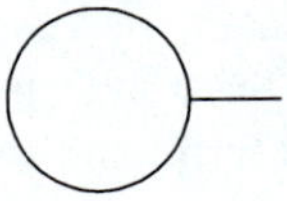

$\widehat{\alpha} = 0.$ (4.78)

Putting all of this together, we can now define the effective action for scalar fermion theories in a unique and frame invariant way. We achieve this using the implicit equation

$$\exp(i\Gamma[\boldsymbol{\Phi}]) = \int \sqrt{|\mathrm{sdet}\, G|}\,[\mathcal{D}\varphi]\exp\left(iS[\varphi] - i\Gamma[\boldsymbol{\Phi}]\frac{\overleftarrow{\delta}}{\delta\Phi^{\widehat{\alpha}}}\Sigma^{\widehat{\alpha}}[\varphi, \boldsymbol{\Phi}]\right). \tag{4.79}$$

The effective action (4.79) can be expanded perturbatively by taking $\Gamma = S + \hbar\Gamma^{(1)} + \hbar^2\Gamma^{(2)} + \cdots$. The one and two loop corrections are then given by

$$\Gamma^{(1)}[\mathbf{\Phi}] = \frac{i}{2} \ln \mathrm{sdet} \left(\overrightarrow{\nabla}^{\hat{\alpha}} S \overleftarrow{\nabla}_{\hat{\beta}} \right), \tag{4.80}$$

$$\begin{aligned} \Gamma^{(2)}[\mathbf{\Phi}] =& \frac{i}{8} S \overleftarrow{\nabla}_{\hat{\alpha}} \overleftarrow{\nabla}_{\hat{\beta}} \overleftarrow{\nabla}_{\hat{\gamma}} \overleftarrow{\nabla}_{\hat{\delta}} {}^{\hat{\delta}\hat{\gamma}}\Delta^{\hat{\beta}\hat{\alpha}}\Delta \\ &+ (-1)^{\hat{\gamma}\hat{\beta}+\hat{\epsilon}(\hat{\delta}+\hat{\beta})} \frac{i}{12} \left(S \overleftarrow{\nabla}_{\hat{\epsilon}} \overleftarrow{\nabla}_{\hat{\gamma}} \overleftarrow{\nabla}_{\hat{\alpha}} \right) {}^{\hat{\alpha}}\Delta^{\hat{\beta}}\, {}^{\hat{\gamma}}\Delta^{\hat{\delta}}\, {}^{\hat{\epsilon}}\Delta^{\hat{\zeta}} \left(\overrightarrow{\nabla}_{\hat{\zeta}} \overrightarrow{\nabla}_{\hat{\delta}} \overrightarrow{\nabla}_{\hat{\beta}} S \right). \end{aligned} \tag{4.81}$$

where ${}^{\hat{\alpha}}\Delta^{\hat{\beta}} = \left(\overrightarrow{\nabla}_{\hat{\alpha}} S \overleftarrow{\nabla}_{\hat{\beta}} \right)^{-1}$ is the propagator.

Notice that both these expressions are reparametrisation invariant as required.

We can also calculate quantum corrections of the theory perturbatively using covariant Feynman diagrams as shown in Sect. 2.5.1. The generalisation of (2.105) in the presence of anticommuting fields is

$= {}^{\hat{\alpha}}\Delta^{\hat{\beta}}$,

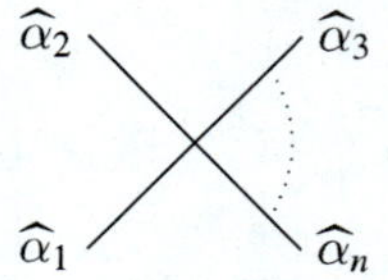

$= \overrightarrow{\nabla}_{\{\hat{\alpha}_1} \ldots \overrightarrow{\nabla}_{\hat{\alpha}_n\}} S.$ (4.82)

Here, the notation $\{\cdots\}$ implies supersymmetrisation over the indices, i.e.

$$\{\alpha_i \cdots \alpha_n\} = \frac{1}{n!} \sum_P (-1)^P P[\alpha_i \cdots \alpha_n], \tag{4.83}$$

where P runs over all permutations of the n indices and $(-1)^P$ gives -1 when the permutation involves an odd number of fermionic commutations and $+1$ otherwise.

By defining these Feynman rules we can also express the two-loop effective action (4.81) graphically in terms of the following graphs:

$\Gamma^{(2)}[\mathbf{\Phi}] =$

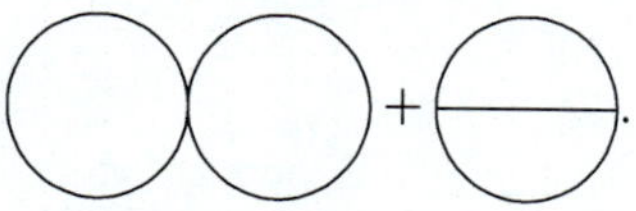

(4.84)

As in the bosonic case, we need only include one particle irreducible graphs in the expansion (4.84). This is because one particle reducible graphs such as

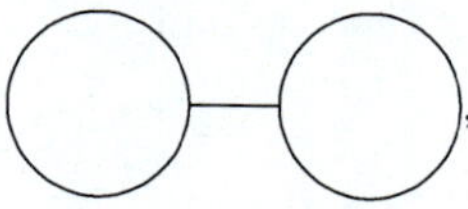

(4.85)

evaluate to zero thanks to (4.78).

4.7 Example: Single Fermion Field

To conclude this chapter let us apply the above formalism to an explicit example. We shall consider a theory with a single scalar field ϕ and a single Dirac fermion ψ. The most general Lagrangian for such a theory with up to two derivatives for the scalar and one derivative for the fermion is

$$\begin{aligned}\mathcal{L} =&\frac{1}{2}k(\phi)\partial_\mu\phi\partial^\mu\phi - \frac{1}{2}h(\phi)\overline{\psi}\gamma^\mu\psi\partial_\mu\phi + \frac{i}{2}g(\phi)\overline{\psi}\gamma^\mu\partial_\mu\psi - \frac{i}{2}g(\phi)\partial_\mu\overline{\psi}\gamma^\mu\psi \\ &- Y(\phi)\overline{\psi}\psi - V(\phi),\end{aligned} \tag{4.86}$$

where k, h, g, Y and V are arbitrary real functions of ϕ.

We can extract the model functions ${}_\alpha k_\beta$, ζ^μ_α and U either by reading them off the Lagrangian (4.86) or by applying the formulae (4.36)–(4.38). Either way we find them to be

$$\begin{aligned}{}_\alpha k_\beta &= \begin{pmatrix} k(\phi) & \mathbf{0}_{1\times 8} \\ \mathbf{0}_{8\times 1} & \mathbf{0}_{8\times 8}\end{pmatrix}, \\ \zeta^\mu_\alpha &= \left(ih(\phi)\overline{\psi}_{\dot{a}}\gamma^\mu_{\dot{a}a}\psi_a,\; g(\phi)\overline{\psi}_{\dot{a}}\gamma^\mu_{\dot{a}a},\; g(\phi)\gamma^\mu_{\dot{a}a}\psi_a\right), \\ U &= Y\overline{\psi}\psi + V.\end{aligned} \tag{4.87}$$

We can then use (4.46) to obtain

$$\zeta_\alpha = \left(ih\overline{\psi}\psi,\; g\overline{\psi}_{\dot{a}},\; g\psi_a\right) \tag{4.88}$$

and can subsequently use (4.50) to find

$${}_\alpha\lambda_\beta = \begin{pmatrix} 0 & \frac{1}{2}(g' - ih)\overline{\psi}_{\dot{a}} & \frac{1}{2}(g' + ih)\psi_a \\ \frac{1}{2}(g' - ih)\overline{\psi}_{\dot{a}} & 0 & g1_4 \\ \frac{1}{2}(g' + ih)\psi_a & g1_4 & 0 \end{pmatrix}, \tag{4.89}$$

where a prime $'$ indicates differentiation with respect to ϕ. Finally, we employ (4.57) and find that

$${}_\alpha\Lambda_\beta = \begin{pmatrix} k & \frac{1}{2}(g' - ih)\overline{\psi}_{\dot{a}} & \frac{1}{2}(g' + ih)\psi_a \\ \frac{1}{2}(g' - ih)\overline{\psi}_{\dot{a}} & 0 & g1_4 \\ \frac{1}{2}(g' + ih)\psi_a & g1_4 & 0 \end{pmatrix}. \tag{4.90}$$

For this theory (4.61) takes the form

$$\begin{pmatrix} k & \frac{1}{2}(g' - ih)\overline{\psi}_{\dot{a}} & \frac{1}{2}(g' + ih)\psi_a \\ \frac{1}{2}(g' - ih)\overline{\psi}_{\dot{a}} & 0 & g1_4 \\ \frac{1}{2}(g' + ih)\psi_a & g1_4 & 0 \end{pmatrix} = {}_\alpha e^a \begin{pmatrix} 1 & 0 & 0 \\ 0 & 0 & 1_4 \\ 0 & 1_4 & 0 \end{pmatrix} {}^b e^{\mathsf{sT}}_\beta. \tag{4.91}$$

This can be solved to give

$$
{}_{\alpha}e^{a}=\begin{pmatrix}\sqrt{k} & \frac{g'+ih}{2\sqrt{g}}\psi_a\, x_{a\dot{a}} & \frac{g'-ih}{2\sqrt{g}}\overline{\psi}_{\dot{a}}\, x^{-1}_{\dot{a}a}\\ 0 & \sqrt{g}x_{a\dot{a}} & 0\\ 0 & 0 & \sqrt{g}x^{-1}_{\dot{a}a}\end{pmatrix}, \tag{4.92}
$$

where $x_{a\dot{a}}$ is an arbitrary invertible 4×4 matrix that can depend on both ϕ and $\overline{\psi}\psi$.

Note that we have chosen the solution (4.92) since it will give the correct signature for the metric. There exists another solution

$$
{}_{\alpha}\widetilde{e}^{a}=\begin{pmatrix}\sqrt{k} & \frac{g'-ih}{2\sqrt{g}}\overline{\psi}_{\dot{a}}x^{-1}_{\dot{a}a} & \frac{g'+ih}{2\sqrt{g}}\psi_a x_{a\dot{a}}\\ 0 & 0 & \sqrt{g}x_{a\dot{a}}\\ 0 & \sqrt{g}x^{-1}_{\dot{a}a} & 0\end{pmatrix}, \tag{4.93}
$$

but this will correspond to a metric with the wrong signature as discussed in Sect. 4.5.

We can now use (4.60) to calculate the field space metric for (4.86) and we find it to be

$$
{}_{\alpha}G_{\beta}=\begin{pmatrix}k-\frac{g'^2+h^2}{2g}\overline{\psi}\psi & -\frac{1}{2}(g'-ih)\overline{\psi}_{\dot{a}} & \frac{1}{2}(g'+ih)\psi_a\\ \frac{1}{2}(g'-ih)\overline{\psi}_{\dot{a}} & 0 & g1_4\\ -\frac{1}{2}(g'+ih)\psi_a & -g1_4 & 0\end{pmatrix}. \tag{4.94}
$$

We see that, as promised, the arbitrary matrix $x_{a\dot{a}}$ does not enter the metric. This metric has superdeterminant

$$
\mathrm{sdet}(G)=\frac{k}{g^8}, \tag{4.95}
$$

which will appear in the path integral measure. Note that, as expected, we have $|\mathrm{sdet}(G)|=|\mathrm{sdet}(\Lambda)|$.

Let us now calculate the connections on the field space using (4.67). We find that the non-zero connections are

$$
\begin{aligned}
{}^{\phi}\Gamma_{\phi\phi}&=\frac{k'}{2k},\\
{}^{\psi_a}\Gamma_{\phi\phi}&=\left[-\frac{h^2+g'^2}{4g^2}+\frac{g''+ih'-\frac{k'}{2k}(g'+ih)}{2g}\right]\psi_a,\\
{}^{\psi_a}\Gamma_{\psi_b\phi}={}^{\psi_a}\Gamma_{\phi\psi_b}&=\frac{g'+ih}{2g}\delta_{ab},\\
{}^{\overline{\psi}_{\dot{a}}}\Gamma_{\phi\phi}&=\left[-\frac{h^2+g'^2}{4g^2}+\frac{g''-ih'-\frac{k'}{2k}(g'-ih)}{2g}\right]\overline{\psi}_{\dot{a}},\\
{}^{\overline{\psi}_{\dot{b}}}\Gamma_{\overline{\psi}_{\dot{a}}\phi}={}^{\overline{\psi}_{\dot{b}}}\Gamma_{\overline{\psi}_{\dot{a}}\phi}&=\frac{g'-ih}{2g}\delta_{\dot{a}\dot{b}},
\end{aligned} \tag{4.96}
$$

and that all others vanish. We can also compute the field-space Riemann tensor using (4.23); however, we find that it vanishes identically with ${}^{\alpha}R_{\beta\gamma\delta} = 0$. We can, therefore, conclude that the field space of the theory (4.86) is flat.

Because the field space is flat it must be possible to transform (4.86) into a canonical theory by performing an appropriate field redefinition. The relevant redefinition is

$$\begin{aligned} \phi \to \widetilde{\phi} &= \int_0^{\phi} \sqrt{k(\phi)}d\phi, \\ \psi \to \widetilde{\psi} &= \sqrt{g(\phi)} \exp\left(\frac{i}{2}\int_0^{\phi} \frac{h(\phi)}{g(\phi)}d\phi\right)\psi, \\ \overline{\psi} \to \widetilde{\overline{\psi}} &= \sqrt{g(\phi)} \exp\left(-\frac{i}{2}\int_0^{\phi} \frac{h(\phi)}{g(\phi)}d\phi\right)\overline{\psi}. \end{aligned} \tag{4.97}$$

Indeed, we find that when expressed in the fields

$$\widetilde{\Phi}^{\alpha} = \left(\widetilde{\phi}, \widetilde{\psi}, \widetilde{\overline{\psi}}\right), \tag{4.98}$$

the Lagrangian (4.86) becomes

$$\mathcal{L} = \frac{1}{2}\partial_{\mu}\widetilde{\phi}\partial^{\mu}\widetilde{\phi} + \frac{i}{2}\widetilde{\overline{\psi}}\gamma^{\mu}\partial_{\mu}\widetilde{\psi} - \frac{i}{2}\partial_{\mu}\widetilde{\overline{\psi}}\gamma^{\mu}\widetilde{\psi} - \widetilde{Y}(\widetilde{\phi})\widetilde{\overline{\psi}}\widetilde{\psi} - \widetilde{V}(\widetilde{\phi}). \tag{4.99}$$

Here, we have defined

$$\begin{aligned} \widetilde{Y}(\widetilde{\phi}) &= g(\phi)Y(\phi), \\ \widetilde{V}(\widetilde{\phi}) &= V(\phi). \end{aligned} \tag{4.100}$$

Notice that, with the choice

$$x_{a\dot{a}} = \exp\left(\frac{i}{2}\int_0^{\phi} \frac{h(\phi)}{g(\phi)}d\phi\right)\delta_{a\dot{a}}, \tag{4.101}$$

we have

$${}_{\alpha}e^{a} = \frac{\overrightarrow{\partial}}{\partial\Phi^{\alpha}}\widetilde{\Phi}^{a}. \tag{4.102}$$

Thus, the field space vielbeins can be written as the Jacobian of a transformation, as expected for a flat supermanifold.

Let us now calculate the one-loop effective potential for (4.99). Expressed in this parametrisation the field space is trivial and, thus, covariant derivatives give identical results to partial derivatives. We, therefore, find that

$$\overrightarrow{\nabla}_\alpha S \overleftarrow{\nabla}_\beta = \begin{pmatrix} -\Box - \widetilde{V}''(\widetilde{\phi}) - \widetilde{\bar{\psi}}\widetilde{\psi}\widetilde{Y}''(\widetilde{\phi}) & -\widetilde{\bar{\psi}}\widetilde{Y}'(\widetilde{\phi}) & \widetilde{\psi}\widetilde{Y}'(\widetilde{\phi}) \\ \widetilde{\bar{\psi}}\widetilde{Y}'(\widetilde{\phi}) & 0 & -\partial\!\!\!/ - \widetilde{Y}(\widetilde{\phi}) \\ -\widetilde{\psi}\widetilde{Y}'(\widetilde{\phi}) & (\partial\!\!\!/ + \widetilde{Y}(\widetilde{\phi}))^T & 0 \end{pmatrix}. \tag{4.103}$$

Plugging this result into (4.80), we see that the one-loop effective action for the theory (4.86) is

$$\begin{aligned}\Gamma[\Phi] =& S[\Phi] - \frac{i}{2}\,\mathrm{Tr}\ln\left(\Box + \widetilde{V}''(\widetilde{\phi}) + \widetilde{\bar{\psi}}\left[2\big(\widetilde{Y}'(\widetilde{\phi})\big)^2(\partial\!\!\!/ + \widetilde{Y})^{-1} - \widetilde{Y}''(\widetilde{\phi})\right]\widetilde{\psi}\right) \\ & - i\,\mathrm{Tr}\ln\big(\partial\!\!\!/ + \widetilde{Y}(\widetilde{\phi})\big). \end{aligned} \tag{4.104}$$

This agrees with previous results in the literature (see e.g. (8.49) in [35]).

References

1. Vilkovisky GA (1984) The unique effective action in quantum field theory. Nucl Phys B234:125–137. (cit. on pp. 50, 51, 55, 56, 63, 64, 67-69, 79, 90, 91, 94, 96, 124, 243)
2. Vilkovisky GA (1984) The gospel according to DeWitt 169–209. (cit. on pp. 50, 51, 55, 56, 63, 64, 67-69, 79, 90, 91, 94, 96, 124, 243)
3. DeWitt BS (1987) The effective action: architecture of fundamental interactions at short distances. In: Proceedings, Les Houches 44th summer school of theoretical physics. Les Houches, France, July 1-August 8, 1985, pt2, pp 1023–1058. (cit. on pp. 67-69, 105, 124, 243)
4. Rebhan A (1987) The vilkovisky-de witt effective action and its application to yang-mills theories. Nucl Phys B288:832–857. (cit. on pp. 69, 105)
5. Helset A, Martin A, Trott M (2020) The geometric standard model effective field theory. JHEP 03:163. arXiv: 2001. 01453 [hep-ph] (cit. on p. 105)
6. Fumagalli J, Postma M, Van Den Bout M (2020) Matching and running sensitivity in non-renormalizable inflationary models. arXiv: 2005.05905 [hep-ph] (cit. on p. 105)
7. Dirac PAM (1928) The quantum theory of the electron. Proc Roy Soc Lond A A117:610–624 (cit. on pp. 105, 112)
8. Fermi E (1934) An attempt of a theory of beta radiation 1. Z Phys 88:161–177 (cit. on p. 105)
9. Schwinger J (1969) Particles and sources (cit. on p. 105)
10. Grassmann H (1845) Neue Theorie der Elektrodynamik. Annalen der Physik 140(1):1–18 (cit. on p. 106)
11. Berezin FA (1966) The method of second quantization. In: Pure and applied physics. Academic Press (cit. on pp. 106, 107, 112)
12. Khalatnikov IM (1955) The Representation of green's function in quantum electrodynamics in the form of continual integrals. JETP 1:568 (cit. on p. 107)
13. Matthews PT, Salam A (1955) Propagators of quantized field. Nuovo Cim 2:120–134 (cit. on p. 107)
14. Martin JL, Kemmer N (1959) The Feynman principle for a Fermi system. In: Proceedings of the royal society of London: series a. Mathematical and physical sciences 251(1267):543–549. https://royalsocietypublishing.org/doi/pdf/10.1098/rspa.1959.0127 (cit. on p. 107)
15. Berezin FA (1987) Introduction to superanalysis. In: Kirillov AA, Leites D (eds) (cit. on p. 107)
16. Kostant B (1977) Graded manifolds, graded Lie theory, and prequantization. In: Differential geometrical methods in mathematical physics. In: Bleuler K, Reetz A (eds) Springer Berlin Heidelberg, Berlin, Heidelberg, pp 177–306 (cit. on p. 107)

17. Batchelor M (1979) Supermanifolds. In: Group theoretical methods in physics. In: Beiglböck W, Böhm A, Takasugi E (eds) Springer Berlin Heidelberg, Berlin, Heidelberg, pp 458–465 (cit. on p. 107)
18. Batchelor M (1979) The structure of supermanifolds. Trans Am Math Soc 253:329–338 (cit. on p. 107)
19. Leites DA (1980) Introduction to the theory of supermanifolds. Russ Math Surv 35(1):1–64 (cit. on p. 107)
20. DeWitt B (1992) Supermanifolds. In: Cambridge monographs on mathematical physics. Cambridge University Press (cit. on pp. 107, 108, 112, 120)
21. Rogers A (2007) Supermanifolds: theory and applications (cit. on pp. 107, 108)
22. Volkov DV, Akulov VP (1973) Is the neutrino a goldstone particle? Phys Lett B 46:109–110 (cit. on p. 107)
23. Wess J, Zumino B (1974) Supergauge transformations in four-dimensions. Nucl Phys B 70. In: Salam A, Sezgin E (eds) pp 39–50 (cit. on pp. 107, 113, 247)
24. Salam A, Strathdee J (1974) Super-gauge transformations. Nucl Phys B 76(3):477–482 (cit. on p. 107)
25. Gates SJ, Grisaru MT, Rocek M, Siegel W (1983) Superspace or one thousand and one lessons in supersymmetry, vol 58. arXiv: hep-th/0108200 (cit. on p. 107)
26. Sohnius MF (1985) Introducing supersymmetry. Phys Rept 128:39–204 (cit. on p. 107)
27. Guajardo Ó (2014) A classical approach to smooth supermanifolds. arXiv:1410.7857. arXiv: 1410.7857 [math.DG] (cit. on p. 107)
28. Khudaverdian H, Voronov T (2020) Thick morphisms of supermanifolds, quantum mechanics, and spinor representation. J Geom Phys 148:103540 (cit. on p. 107)
29. Kellett M, Morris TR (2020) Parisi-Sourlas supergravity. Class Quant Grav 37(19):195018. arXiv: 2005.03400 [hep-th] (cit. on p. 107)
30. Yukawa H (1935) On the interaction of elementary particles I. Proc Phys Math Soc Jap 17:48–57 (cit. on p. 116)
31. Bjorken JD, Drell SD (1965) Relativistic quantum fields. In: International series in pure and applied physics. McGraw-Hill (cit. on p. 117)
32. Tomonaga S (1946) On a relativistically invariant formulation of the quantum theory of wave fields. Prog Theor Phys 1:27–42 (cit. on p. 118)
33. Schwinger JS (1948) Quantum electrodynamics: i a covariant formulation. Phys Rev 74. In: Milton KA (ed) p. 1439 (cit. on p. 118)
34. Feynman RP (1949) Space - time approach to quantum electrodynamics. Phys Rev 76. In: Brown LM (ed) pp. 769–789 (cit. on p. 118)
35. Zinn-Justin J (1989) Quantum field theory and critical phenomena, vol 77, p 198 (cit. on p. 130)

Chapter 5
The Eisenhart Lift

Nothing is real.
—John Lennon

The potential term did not play a major role in the geometric description of QFTs developed in the previous chapters. The potential was not included in the construction of the field space, but was instead considered an external force, separate from the field-space manifold.

Because of this, the connection between the physics of the field theory and the geometry of the field space is weakened. This is especially true for scalar field theories, for which the existence of a potential term breaks the one-to-one correspondence between Noether symmetries of the field theory and Killing vectors of the field space as a well as the correspondence between the equations of motion of the field theory and the geodesic equation of the field space (see Sect. 2.2.1).

In this chapter we address these issues by showing how the potential term can also be described geometrically. We achieve this by employing a technique from classical mechanics known as the Eisenhart Lift [5–11], which allows conservative forces to be explained as the consequence of geometry.

5.1 The Eisenhart Lift in Classical Mechanics

Consider a classical particle of mass m, moving in D spatial dimensions subject to a conservative force

$$\boldsymbol{F} = -\nabla V, \tag{5.1}$$

where V is the potential. Such a system can be described by the Lagrangian

K. Finn, *Geometric Approaches to Quantum Field Theory*, Springer Theses,
https://doi.org/10.1007/978-3-030-85269-6_5

$$L = \frac{1}{2} m \sum_{i=1}^{D} \dot{x}_i^2 \; - \; V(\boldsymbol{x}) \,, \tag{5.2}$$

where x^i with $i \in \{1, \ldots, D\}$ (collectively $\boldsymbol{x}$) are the spatial coordinates and the overdot $\dot{}$ denotes differentiation with respect to time t.

We can solve the Euler-Lagrange equations for (5.2) to find the following equations of motion

$$m\ddot{x}_i \; = \; -\frac{\partial V(\boldsymbol{x})}{\partial x^i} = F_i \,. \tag{5.3}$$

As expected, this is simply Newton's second law [12].

In Eq. (5.3) the force $\boldsymbol{F}$ is disconnected from the geometry of the space the particle lives in (flat, Euclidean space in this case). However, Eisenhart [5] showed that this force can actually be reinterpreted in a geometric way. If we consider the particle to be moving in a space with a hidden extra dimension and choose the metric of this extended space appropriately, the effect of the force $\boldsymbol{F}$ can be fully recreated by the curvature of space. Below we repeat the derivation of this discovery.

Let us extend the space in which this particle is moving by adding a new coordinate y. We take the metric of this new, extended space to be

$$g_{IJ} \; = \; \begin{pmatrix} \delta_{ij} & 0 \\ 0 & \dfrac{M^2}{mV(\boldsymbol{x})} \end{pmatrix} , \tag{5.4}$$

where M is an arbitrary mass scale[1] introduced to keep dimensions consistent. We consider a free particle in this space, which has the Lagrangian

$$L = \frac{1}{2} m g_{IJ} \dot{X}^I \dot{X}^J = \frac{1}{2} m \sum_{i=1}^{n} \dot{x}_i^2 \; + \; \frac{1}{2} \frac{M^2}{V(\boldsymbol{x})} \dot{y}^2 , \tag{5.5}$$

where $X^I = \{x^i, y\}$ are the coordinates on the extended space.

Since this particle is free it will follow a geodesic of the extended space. Indeed, the equations of motion for the Lagrangian (5.5) give

$$\ddot{X}^I + \Gamma^I_{JK} \dot{X}^J \dot{X}^K = 0, \tag{5.6}$$

where

$$\Gamma^I_{JK} = \frac{1}{2} g^{IL} \left(\frac{\partial g_{LK}}{\partial X^J} + \frac{\partial g_{JL}}{\partial X^K} - \frac{\partial g_{JK}}{\partial X^L} \right) \tag{5.7}$$

are the Christoffel symbols for the metric (5.4).

---

[1] Technically M has units of mass $\times$ speed. However, we are still working in units where $c = 1$ and so M can be considered a mass.

The non-zero Christoffel symbols are

$$\Gamma^{x_i}_{yy} = \frac{M^2}{2mV}\frac{\partial V}{\partial x_i}, \qquad \Gamma^{y}_{yx_i} = -\frac{1}{2V}\frac{\partial V}{\partial x_i}. \tag{5.8}$$

We can, therefore, write (5.6) more explicitly to show the dependence on the new coordinate y and we find

$$m\ddot{x}_i = -\frac{1}{2}\frac{M^2}{V^2}\frac{\partial V}{\partial x^i}\,\dot{y}^2\,, \tag{5.9}$$

$$M\frac{d}{dt}\left(\frac{M\dot{y}}{V(\boldsymbol{x})}\right) = 0\,. \tag{5.10}$$

Let us examine these equations more closely, starting with (5.10). The solution to this equation is clearly

$$\frac{M\dot{y}}{V(\boldsymbol{x})} = A, \tag{5.11}$$

where A is a constant. We can plug this condition into (5.9) to find

$$m\ddot{x}_i = -\frac{A^2}{2}\frac{\partial V}{\partial x^i}. \tag{5.12}$$

We notice that if we choose initial conditions for which $A^2 = 2$ then (5.12) becomes identical to (5.3). Therefore, if we are ignorant to the existence of the dimension y, then this particle will appear to us as though it were experiencing the external force (5.1).

Notice that both (5.9) and (5.10) are invariant under constant rescalings of time

$$t \to \tilde{t} = ct, \tag{5.13}$$

where c is a constant. This invariance has a geometric interpretation. For a free theory, the particle must follow a geodesic. But there are an infinite number of ways to parametrise such a geodesic, corresponding to the infinitely many choices of affine parameter. The invariance under (5.13) expresses the freedom we have to choose which affine parameter should be associated with time.

The need for the condition $A^2 = 2$ is closely related to the freedom of affine parametrisation. The original equations of motion (5.3) do not have that freedom and so it makes sense that they will only agree for a particular choice of affine parameter. In addition, by adding a new dimension to the space we require an additional piece of information to set the initial conditions of the system.[2] We, therefore, need an

[2] Technically we require two additional pieces of information, but the invariance of (5.5) under shifts in y means that the initial value of the y coordinate is irrelevant.

additional boundary condition in order to fully constrain the system. This boundary condition is the requirement $A^2 = 2$.

The Eisenhart lift shows that any conservative force in classical mechanics can be considered *fictitious*—arising only because we have neglected some aspect of the system. The system can be perfectly well described as a free particle moving in the higher dimensional space, but by neglecting the extra dimension y, we must introduce the force $\boldsymbol{F}$ to account for its motion.

This is analogous to the centrifugal and Coriolis forces that arise in rotating reference frames such as the surface of the Earth [13]. By neglecting the non-inertial nature of such frames we are forced to introduce new fictitious forces to account for the motion.

Another example of a fictitious force is gravity. Einstein showed that the spacetime manifold surrounding a massive object is inherently curved [1]. If we take this curvature into account then objects simply follow geodesics of the spacetime and there is no need to introduce a gravitational force. However, if we ignore the curvature and insist on treating space as though it were flat, we have to introduce a force to explain an object's deviation from a straight line. At leading order the required force satisfies precisely Newton's law of gravitation [12].

We should contrast the Eisenhart lift to the work of Kaluza [3] and Klein [4] (see [14] for a review). In Kaluza–Klein theory, the spacetime manifold was extended by an additional dimension and it was shown that observers ignorant of the extra dimension would experience a fictitious electromagnetic force. However, in the case of Kaluza–Klein theory, the metric of the extended spacetime is considered to be dynamical and the fictitious force arises due to the additional degrees of freedom present in the extended metric, which obey Maxwell's equations. In contrast, in the Eisenhart lift formalism, the metric is taken to be fixed and, therefore, the fictitious force arises only due to the additional curvature present in the extended space.

5.2 Quantum Mechanics

We have so far shown the equivalence between the original system with a force and the lifted geometric system only at the classical level. In this section we will investigate the quantum relationship between the two systems.

In order to achieve this, it will prove useful to reformulate the above derivation in the Hamiltonian formalism [15, 16]. We therefore define the Hamiltonian

$$H = P_I \dot{X}^I - L, \tag{5.14}$$

in the usual way. Here

$$P_I \equiv \frac{\partial L}{\partial \dot{X}^I} \tag{5.15}$$

is the canonical momentum conjugate to X^I. We shall refer to coordinates X^I and momenta P_I collectively as $\boldsymbol{X}$ and $\boldsymbol{P}$, respectively.

Applying these definitions to the lifted, geometric system described by (5.5) we find that the canonical momenta are given by

$$P_I = mg_{IJ}\dot{X}^J = \{p_i, p_y\}, \tag{5.16}$$

where

$$p_i = m\dot{x}^i, \tag{5.17}$$

$$p_y = \frac{M^2}{V(\boldsymbol{x})}\dot{y}. \tag{5.18}$$

Meanwhile, the Hamiltonian is

$$H = \sum_{i=1}^{D} \frac{p_i^2}{2m} + \frac{V(\boldsymbol{x})}{M^2}\frac{p_y^2}{2}. \tag{5.19}$$

We can now apply Hamilton's equations,

$$\dot{X}^I = \frac{\partial H}{\partial P_I}, \qquad \dot{P}_I = \frac{\partial H}{\partial X_I}, \tag{5.20}$$

to (5.19) which gives

$$\dot{x}_i = p_i/m, \qquad \dot{p}_i = -\frac{p_y^2}{2M^2}\frac{\partial V}{\partial x^i}, \tag{5.21}$$

$$\dot{y} = p_y V/M^2, \qquad \dot{p}_y = 0, \tag{5.22}$$

as the equations of motion.

From (5.22) we see that p_y is a constant of the motion. In fact, by comparing to (5.18)–(5.11) we see that

$$p_y = AM, \tag{5.23}$$

where A is defined in (5.11). Plugging this result into (5.21) yields

$$\dot{p}_i = -\frac{A^2}{2}\frac{\partial V}{\partial x^i}, \tag{5.24}$$

which is identical to (5.12). Therefore, when $A^2 = 2$, or equivalently when $p_y^2 = 2M^2$, we recover the equations of motion for the original system described by (5.2).

Let us now quantise this system in the canonical way, by promoting the coordinates X^I and canonical momenta P_I to the operators $\widehat{X}^I$ and $\widehat{P}_I$, respectively [17, 18]. As usual, we shall infer the commutation relations of our newly defined operators from the Poisson brackets of their classical analogue so that

$$
\begin{aligned}
\left[\widehat{X}^I, \widehat{P}_J\right] &= i\left\{X^I, P_J\right\}, \\
\left[\widehat{X}^I, \widehat{X}^J\right] &= i\left\{X^I, X^J\right\}, \\
\left[\widehat{P}_I, \widehat{P}_J\right] &= i\left\{P_I, P_J\right\}.
\end{aligned} \tag{5.25}
$$

Here $[\cdots]$ denotes the commutator and $\{\cdots\}$ denotes the Poisson bracket.

We recall that the Poisson bracket is defined as

$$
\{f, g\} \equiv \frac{\partial f}{\partial X^I}\frac{\partial g}{\partial P_I} - \frac{\partial g}{\partial X^I}\frac{\partial f}{\partial P_I} \tag{5.26}
$$

for generic functions of the canonical coordinates $f(\boldsymbol{X}, \boldsymbol{P})$ and $g(\boldsymbol{X}, \boldsymbol{P})$. Therefore, (5.25) becomes

$$
\begin{aligned}
\left[\widehat{X}^I, \widehat{P}_J\right] &= i\delta_I^J, \\
\left[\widehat{X}^I, \widehat{X}^J\right] = \left[\widehat{P}_I, \widehat{P}_J\right] &= 0.
\end{aligned} \tag{5.27}
$$

Separating out the fictitious dimension y, we see that

$$
[\hat{x}^i, \hat{p}_j] = i\delta^i_j, \qquad\qquad [\hat{y}, \hat{p}_y] = i, \tag{5.28}
$$

with all other commutators vanishing.

Notice that anything physically observable will necessarily depend on $\hat{x}^i$ and $\hat{p}_i$ only and will, therefore, commute with with any quantity that depends only on the fictitious coordinates $\hat{y}$ and $\hat{p}_y$. This is important because it means that the fictitious subspace of the lifted manifold cannot be probed by observers living in the original space and so such observers will be ignorant to the existence of these extra dimensions.

We now derive the time independent Schrödinger equation (TISE) for the above system, which is given by [19, 20]

$$
\widehat{H}\,|\Psi\rangle = E\,|\Psi\rangle\,. \tag{5.29}
$$

Here $\widehat{H} = H(\widehat{\boldsymbol{X}}, \widehat{\boldsymbol{P}})$ is the Hamiltonian operator, $|\Psi\rangle$ is the $D+1$ dimensional energy eigenstate of the system and E is its associated eigenvalue. For the lifted system with Hamiltonian (5.19), the TISE is

$$
-\frac{1}{2m}\sum_{i=1}^{D}\frac{\partial^2\Psi(\boldsymbol{x}, y)}{\partial x^{i\,2}} - \frac{V(\boldsymbol{x})}{2M^2}\frac{\partial^2\Psi(\boldsymbol{x}, y)}{\partial y^2} = E\Psi(\boldsymbol{x}, y). \tag{5.30}
$$

Here $\Psi(\boldsymbol{x}, y) = \langle \boldsymbol{x}, y|\Psi\rangle$ is the wavefunction expressed in the position basis, and we have used the fact that

$$\langle \boldsymbol{x}, y| P_I |\Psi\rangle = -i\frac{\partial \Psi(\boldsymbol{x}, y)}{\partial X^I}. \tag{5.31}$$

Let us now attempt to solve (5.30) by assuming a separable solution

$$\Psi(\boldsymbol{x}, y) = \psi(\boldsymbol{x})\xi(y). \tag{5.32}$$

With this assumption, (5.30) becomes

$$\left(\frac{1}{2m}\sum_{i=1}^{D}\frac{\partial^2\psi}{\partial x^{i\,2}} + E\psi\right)\frac{M^2}{V(\boldsymbol{x})\psi} = -\frac{1}{2\xi}\frac{\partial^2\xi}{\partial y^2} = \frac{1}{2}p^2, \tag{5.33}$$

where p is a constant.

We see that the solution for $\xi(y)$ is a pure momentum mode, i.e.

$$\xi(y) = e^{ipy}. \tag{5.34}$$

Therefore, p can be interpreted as the eigenvalue of $\hat{p}_y$. Notice that $[\widehat{H}, \hat{p}_y] = 0$ and thus p is a good quantum number and is conserved as expected by the equations of motion.

However, there is a problem with the solution (5.34) – it cannot be properly normalised if y is allowed to run over the entire real line. This motivates us to consider y to be a compactified dimension, restricted to lie in the range $y \in [0, \ell)$, for some compactification length ℓ. In addition to allowing the wavefunction to be normalised, this also restricts the momentum p to take on one of a discrete set of values given by

$$p = \frac{2\pi k}{\ell}, \tag{5.35}$$

where $k \in \mathbb{Z}$ is an integer quantum number.

Substituting (5.34) and (5.35) into (5.33) we find

$$-\frac{1}{2m}\sum_{i=1}^{D}\frac{\partial^2\psi}{\partial x^{i\,2}} + \frac{2\pi^2k^2}{\ell^2M^2}V(\boldsymbol{x})\psi = E\psi. \tag{5.36}$$

We notice that for the ground state, $k = 0$, (5.36) corresponds the TISE for a *free* particle on the original manifold. Unless this manifold is compact (which is usually not the case) then the corresponding plane-wave solutions are not normalisable. We therefore conclude that the lowest-energy normalisable eigenstate for this system corresponds to $k = \pm 1$.

Since we are free to choose the compactification length, we can set it to

$$\ell = \frac{\sqrt{2}\pi}{M}. \tag{5.37}$$

With this choice, the lowest-energy state satisfies the Eisenhart condition $p^2 = 2M^2$. The TISE for the lifted system (5.36) also takes on the simpler form

$$-\frac{1}{2m}\sum_{i=1}^{D}\frac{\partial^2\psi}{\partial x^{i\,2}} + k^2\,V(\boldsymbol{x})\psi = E\psi. \tag{5.38}$$

We see that if $k = \pm 1$ this equation reduces exactly to the TISE for the original D-dimensional system with potential $V(\boldsymbol{x})$. Thus, in this case $\psi(\boldsymbol{x})$ will be one of the original energy eigenstates. However, for states with $k \neq \pm 1$, we see that the system behaves as though the potential were scaled by a constant amount.

In the classical case, we could account for this rescaling by taking advantage of the freedom of affine parametrisation to rescale our time parameter according to (5.13). We notice, by considering the time dependent Schrödinger equation [21, 22]

$$\widehat{H}\,|\Psi(t)\rangle = i\frac{\partial}{\partial t}\,|\Psi(t)\rangle\,, \tag{5.39}$$

that rescaling time is equivalent to rescaling all energy levels by a constant amount.

However, in the quantum case, a simple rescaling of time is not enough to transform (5.38) into the TISE for the original system. Instead we must rescale both the energy levels *and* the mass of the particle

$$m_k = k^2 m, \qquad\qquad E_k = \frac{E}{k^2}. \tag{5.40}$$

At the classical level, such a transformation is indistinguishable from (5.13), but at the quantum level, the spatial extent of the wavefunction allows us to distinguish between these two transformations.

Nonetheless, if we do transform both the energy levels and the mass in this way we obtain

$$-\frac{1}{2m_k}\sum_{i=1}^{D}\frac{\partial^2\psi}{\partial x^{i\,2}} + V(\boldsymbol{x})\psi = E_k\psi, \tag{5.41}$$

which is precisely the TISE for the original system. We therefore see that, while less trivially realised than in the classical case, the equivalence between the original system described by (5.2) and the lifted system described by (5.5) does still persist at the quantum level.

5.3 Example: Simple Harmonic Oscillator

As an example let us consider the simple harmonic oscillator. Such a system depends on a single coordinate x and has a quadratic potential $V(x) = \frac{1}{2}m\omega^2 x^2$, where ω is the frequency. The Lagrangian for the system is therefore

$$L = \frac{1}{2}m\dot{x}^2 - \frac{1}{2}m\omega^2 x^2 \tag{5.42}$$

and the equation of motion is

$$\ddot{x} = -\omega^2 x \tag{5.43}$$

with solution

$$x = x_0 \cos\left[\omega(t - t_0)\right]. \tag{5.44}$$

Here, x_0 and t_0 are parameters set by the initial conditions.

Let us now apply the Eisenhart lift to this system. We, therefore, extend the space with an additional dimension y and equip the new, two-dimensional space with a metric

$$g_{IJ} = \begin{pmatrix} 1 & 0 \\ 0 & \frac{2M^2}{m^2\omega^2 x^2} \end{pmatrix}. \tag{5.45}$$

We then consider a free particle moving in this two dimensional space with a Lagrangian

$$L = \frac{1}{2}m g_{IJ}\dot{X}^I\dot{X}^J = \frac{1}{2}m\dot{x}^2 + \frac{M^2}{m\omega^2 x^2}\dot{y}^2 , \tag{5.46}$$

where $X^I = \{x, y\}$.

We can apply the Euler Lagrange equations to (5.46) to obtain the geodesic equation for this free particle. We find

$$m\ddot{x} = -\frac{2M^2}{m\omega^2 x^3}\dot{y}^2 , \tag{5.47}$$

$$M\frac{d}{dt}\left(\frac{2M\dot{y}}{m\omega^2 x^2}\right) = 0 . \tag{5.48}$$

Equation (5.48) implies

$$\dot{y} = A\frac{m\omega^2}{2M}x^2, \tag{5.49}$$

where A is a constant. We can plug this into (5.47) to find

$$\ddot{x} = -\frac{A^2}{2}\omega^2 x , \tag{5.50}$$

which can be solved to give

$$x(t) = x_0 \cos\left[\frac{A}{\sqrt{2}}\omega(t - t_0)\right]. \tag{5.51}$$

As expected, the particle undergoes simple harmonic motion in the x direction.

Finally, we can plug (5.51) into (5.49) to find

$$\dot{y} = A\frac{m\omega^2 x_0^2}{2M}\cos^2\left[\frac{A}{\sqrt{2}}\omega(t-t_0)\right], \tag{5.52}$$

which has the solution

$$y(t) = y_0 + A\frac{m\omega^2 x_0^2}{4M}(t-t_0) + \frac{m\omega x_0^2}{4\sqrt{2}M}\sin\left[\sqrt{2}A\omega(t-t_0)\right], \tag{5.53}$$

where y_0 is an irrelevant constant of integration.

The solutions (5.51) and (5.53) have four free parameters. The parameters x_0 and t_0 set the amplitude and phase, respectively, of the harmonic motion as before. The parameter y_0 gives the initial y position, which is irrelevant since the system is shift-symmetric in y.

Finally, there is the parameter A, which has been discussed before. We see that A appears in (5.51) and (5.53) only as a coefficient of $t - t_0$. Therefore, the value of A can be changed by rescaling the time coordinate t as shown previously. In particular, there is always a choice of time coordinate $\tau = At/\sqrt{2}$ for which (5.51) reduces to (5.44). For any other choice, the free particle will still emulate the system (5.42) and will still undergo simple harmonic motion in the x direction, but will appear in fast forward (if $A^2 > 2$) or slow motion (if $A^2 < 2$) compared to (5.44).

In Fig. 5.1 we plot the solutions (5.51) and (5.53) taking $m = 1\,\text{kg}$, $\omega = 2\pi\,\text{s}^{-1}$ and $M = 1\,\text{kg m s}^{-1}$. Note that, since we are temporarily using SI units, $c \neq 1$ and so the units of M are mass$\times$speed. We fix $t_0 = y_0 = 0$ for all trajectories and consider two amplitudes $x_0 = 1\,\text{m}$ (solid red) and $x_0 = 1.5\,\text{m}$ (dashed blue). In addition we consider two values of the parameter A: $A = \sqrt{2}$ and $A = 5\sqrt{2}$.

As expected, the choice of A does not affect the trajectory in the $x - y$ plane, only how fast this trajectory is traversed. We, therefore, show this difference by plotting markers at equal time intervals of $\Delta t = 0.05\,\text{s}$ using crosses for $A = \sqrt{2}$ and squares for $A = 5\sqrt{2}$. Notice that when $A = 5\sqrt{2}$, the trajectory is traversed five times faster than when $A = \sqrt{2}$.

We see that, regardless of the value of x_0, there is 0.5s between peaks when $A = \sqrt{2}$ and 0.1s when $A = 5\sqrt{2}$. Therefore, the period of oscillation is independent of the amplitude as expected.

For an observer unaware of the dimension y, the motion projected onto the x coordinate is simple harmonic. Such observers would need to introduce a force $F = -m\omega^2 x$ to account for this motion. However, if we consider the full two-dimensional space we see that this force is fictitious and the particle is merely following a geodesic of the curved space.

Let us now consider the quantum aspects of this system. The original system has a Hamiltonian operator given by

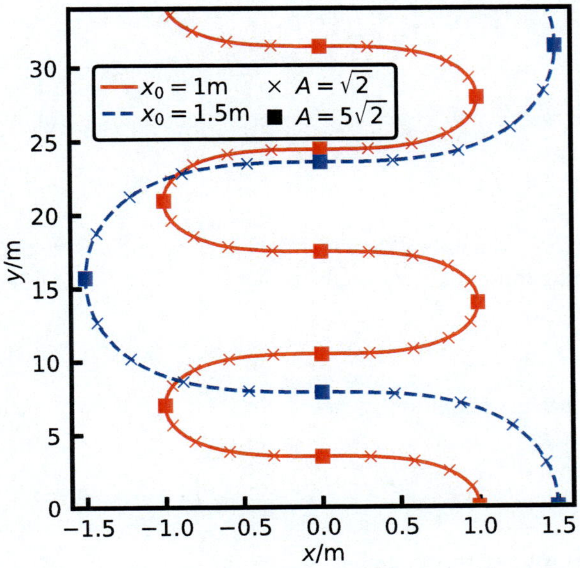

Fig. 5.1 Motion of a free particle in the two-dimensional space with metric (5.45) and angular frequency $\omega = 2\pi\,\mathrm{s}^{-1}$. The solid red line and dashed blue line indicate trajectories with $x_0 = 1$ m and $x_0 = 1.5$ m, respectively. The markers are placed at equal time intervals of $\Delta t = 0.05$ s with crosses for $A = \sqrt{2}$ and squares for $A = 5\sqrt{2}$. Notice how varying A does not change the trajectory, only how quickly it is traversed. Figure reproduced from [37]. © 2018 American Physical Society

$$\widehat{H} = \frac{\hat{p}_x^2}{2m} + \frac{1}{2}m\omega^2\hat{x}^2, \tag{5.54}$$

where $\hat{x}$ and $\hat{p}_x$ are the operators corresponding to the coordinate x and its conjugate momentum p_x, respectively. This leads to the following TISE:

$$-\frac{1}{2m}\frac{\partial^2\psi}{\partial x^2} + \frac{1}{2}m\omega^2x^2\psi = E\psi. \tag{5.55}$$

As is well known (see e.g. [23, 24]) this equation yields a discrete spectrum of energy eigenstates $|n\rangle$, which in the position basis are given by

$$\psi_n(x) \equiv \langle\hat{x}|n\rangle = \frac{1}{\sqrt{2^n n!}}\left(\frac{m\omega}{\pi}\right)^{\frac{1}{4}} e^{-\frac{1}{2}m\omega x^2} H_n\left(\sqrt{m\omega}x\right). \tag{5.56}$$

Here H_n is the nth Hermite polynomial [25] and $n \in \mathbb{Z}$ is the energy level.

This system is most easily analysed using the ladder operators [18]

$$\hat{a} = \sqrt{\frac{m\omega}{2}}\left(\hat{x} + \frac{i}{m\omega}\hat{p}_x\right), \tag{5.57}$$

$$\hat{a}^{\dagger} = \sqrt{\frac{m\omega}{2}} \left(\hat{x} - \frac{i}{m\omega} \hat{p}_x \right). \tag{5.58}$$

Defined in this way, $\hat{a}$ and $\hat{a}^{\dagger}$ act on the energy eigenstates as follows:

$$\hat{a} |n\rangle = \sqrt{n} |n-1\rangle , \qquad \hat{a}^{\dagger} |n\rangle = \sqrt{n+1} |n+1\rangle . \tag{5.59}$$

They are, therefore, sometimes known as the lowering and raising operators, respectively. Combining these properties tells us that

$$\hat{a}^{\dagger}\hat{a} |n\rangle = n |n\rangle . \tag{5.60}$$

We notice that the Hamiltonian operator (5.54) can be written in terms of the ladder operators as

$$\widehat{H} = \omega \left(\hat{a}^{\dagger}\hat{a} + \frac{1}{2} \right). \tag{5.61}$$

The energy spectrum of this system is, therefore,

$$E_n = \langle n|\widehat{H}|n\rangle = \omega \left(n + \frac{1}{2} \right). \tag{5.62}$$

Now let us considered the lifted system with Lagrangian (5.46). The corresponding Hamiltonian operator is given by

$$\widehat{H} = \frac{\hat{p}_x^2}{2m} + \frac{m\omega^2\hat{x}^2}{2M^2} \hat{p}_y^2, \tag{5.63}$$

where $\hat{p}_y$ is the momentum operator conjugate to the fictitious coordinate y.

Our work in the previous section, tells us that the eigenstates of this system will be given, in the position basis, by

$$\Psi_{k,n}(x, y) \equiv \langle x, y|k, n\rangle = \psi_{n,k}(x) \exp\left(\frac{2\pi i k y}{\ell} \right), \tag{5.64}$$

where k and n are the two quantum numbers and

$$\psi_{n,k}(x) \equiv \psi_n(x)|_{m\to k^2 m} = \psi_n(kx) \tag{5.65}$$

with ψ_n defined in (5.56). As before, we have compactified the y direction with a compactification radius satisfying (5.37).

We can construct similar ladder operators for this system. These are defined by

$$\hat{\mathfrak{a}} \equiv \sqrt{\frac{m\omega}{2}} \left(\frac{2\hat{x}\hat{p}_y}{M} + \frac{i\hat{p}_x}{m\omega} \right), \tag{5.66}$$

$$\hat{a}^{\dagger} \equiv \sqrt{\frac{m\omega}{2}} \left(\frac{2\hat{x}\hat{p}_y}{M} - \frac{i\hat{p}_x}{m\omega} \right) . \tag{5.67}$$

Such definitions allow us to write the Hamiltonian (5.63) as

$$\widehat{H} = \omega \left(\hat{a}^{\dagger}\hat{a} + \frac{\hat{p}_y}{2\sqrt{2}M} \right) . \tag{5.68}$$

The new ladder operators act on the eigenstates (5.64) as follows:

$$\hat{a}\,|k, n\rangle = \sqrt{kn}\,|k, n-1\rangle , \qquad \hat{a}^{\dagger}\,|k, n\rangle = \sqrt{k(n+1)}\,|k, n+1\rangle , \tag{5.69}$$

and so satisfy

$$\hat{a}^{\dagger}\hat{a}\,|k, n\rangle = kn\,|k, n\rangle . \tag{5.70}$$

We therefore see that

$$\widehat{H}\,|k, n\rangle = k\omega \left(n + \frac{1}{2} \right) |k, n\rangle \tag{5.71}$$

and, hence, the energy levels of the lifted system are

$$E_{k,n} = kE_n = \frac{1}{k^2}\, E_n|_{m\to k^2 m} \tag{5.72}$$

as expected.

5.4 One-Dimensional Field Theories

Now that we have seen how the Eisenhart lift works in classical and quantum mechanics, let us apply the same technique to QFT. As we have seen in the previous chapters, such theories have a geometric description in terms of the field space. The question is, can the potential term in a QFT be removed by extending the field space in the same way that a conservative force can be removed in classical mechanics by extending physical space?

We start by considering field theories in one dimension, or equivalently, homogeneous field theories. We consider a theory of N scalar fields ϕ^A (collectively ϕ) and a Lagrangian

$$L = \frac{1}{2} k_{AB}(\phi)\dot{\phi}^A\dot{\phi}^B - V(\phi). \tag{5.73}$$

The field space for such a theory is an N-dimensional manifold with metric k_{AB}.

As discussed in Sect. 2.2.1 there is a strong connection between the physics of (5.73) and the geometry of this field space. For example, we saw how there is a one-to-one correspondence between Noether symmetries of (5.73) and Killing vectors of the field space, provided such vectors also leave the potential invariant. We also saw that the equations of motion for (5.73) are strongly related to the geodesic equation on the field space. Indeed, applying the Euler Lagrange equations to (5.73) gives

$$\ddot{\phi}^A + \Gamma^A_{BC}\dot{\phi}^B\dot{\phi}^C = -k^{AB}\partial_B V, \tag{5.74}$$

where

$$\Gamma^A_{BC} = \frac{1}{2}k^{AD}\left(\partial_B k_C + \partial_C k_{BD} - \partial_D k_{BC}\right) \tag{5.75}$$

are the field-space Christoffel symbols, k^{AB} is the inverse of k_{AB}, and $\partial_A \equiv \partial/\partial\phi^A$.

We can see the reason for this connection by comparing (5.73) with (5.2). Indeed, other than relabelling $x^i \to \phi^A$ and setting the mass to $m = 1$, the only difference between these two Lagrangians is that (5.2) assumed a Euclidean space $g_{ij} = \delta_{ij}$, whereas (5.73) allows for a more general field space metric k_{AB}. It, therefore, seems that the Eisenhart lift should apply just as readily to these theories.

Let us, therefore, proceed analogously to Sect. 5.1. We shall extend the field space manifold by introducing a new scalar field χ. We shall take coordinates $\Phi^I = \{\phi^A, \chi\}$ on the extended manifold with $I \in \{1, \ldots, N+1\}$ and equip it with a metric

$$G_{IJ} \equiv \begin{pmatrix} k_{AB}(\phi) & 0 \\ 0 & \dfrac{M^4}{V(\phi)} \end{pmatrix}, \tag{5.76}$$

where M is again an arbitrary mass scale. We use this extended manifold to construct a purely kinetic Lagrangian

$$L = \frac{1}{2}G_{IJ}\,\dot{\Phi}^I\dot{\Phi}^J = \frac{1}{2}k_{AB}(\phi)\,\dot{\phi}^A\dot{\phi}^B + \frac{1}{2}\frac{M^4}{V(\phi)}\dot{\chi}^2\,. \tag{5.77}$$

We note that the new field χ has a kinetic term and is, thus, fully dynamical. It is, therefore, incorrect to call χ an auxiliary field and we instead call it *fictitious*.

The Lagrangian (5.77) has no potential term and so the relation between the equations of motion of the fields and the geodesic equation for (5.76) is exact. Indeed, applying the Euler Lagrange equation to (5.77) yields

$$\ddot{\Phi}^I + \Gamma^I_{JK}\dot{\Phi}^J\dot{\Phi}^K = 0, \tag{5.78}$$

where

$$\Gamma^I_{JK} = \frac{1}{2}G^{IL}\left(\partial_J G_{LK} + \partial_K G_{JL} - \partial_L G_{JK}\right) \tag{5.79}$$

are the Christoffel symbols of the extended field space.

We can separate out the new field χ and rewrite (5.78) as

$$\ddot{\phi}^A + \Gamma^A_{BC}\dot{\phi}^B\dot{\phi}^C + \frac{1}{2}M^4 k^{AB}\frac{\partial_B V}{V^2}\dot{\chi}^2 = 0, \tag{5.80}$$

$$\frac{d}{dt}\left(\frac{M^2\dot{\chi}}{V(\phi)}\right) = 0. \tag{5.81}$$

Equation (5.81) tells us that

$$A = \frac{M^2\dot{\chi}}{V(\phi)} \tag{5.82}$$

is a constant. Plugging this knowledge into (5.80) we get

$$\ddot{\phi}^A + \Gamma^A_{BC}\dot{\phi}^B\dot{\phi}^C = -\frac{A^2}{2}k^{AB}\partial_B V. \tag{5.83}$$

If $A^2 = 2$ then this is identical to (5.74).

Provided this condition is satisfied, our new purely kinetic theory (5.77) is equivalent to the original theory with a potential (5.73). We have, therefore, succeeded in describing this potential term geometrically and including it in the definition of the field space.

We note that, as in the previous section, the value of A can be rescaled by redefining the notion of time using (5.13), which is a symmetry of the equations of motion (5.78). By performing an appropriate redefinition of time, the condition $A^2 = 2$ can always be met.

In the case of field theories, however, we can make a stronger statement. In the theory we are considering, the scalar fields ϕ^A are taken to be the only ones in existence. Therefore, any clock or other time-keeping device will be necessarily made out of these fields. But, a reparametrisation of the form (5.13) causes the evolution of *all* fields to slow down or speed up by the same amount. Any clock will, therefore, slow down or speed up by the same factor as whatever it is measuring and no relative difference will be observed. The value of A can never be measured by an observer that only has access to the scalar fields ϕ^A and it is, therefore, unphysical. The condition $A^2 = 2$ is only needed when comparing to theories that do not have this freedom, such as (5.73).

The benefit of incorporating the potential term into the field space is a much closer relationship between the physics of the field theory and the geometry of the field space. We have already seen that the equations of motion for (5.77) are equivalent to the geodesic equations for (5.76) with no need for an external force.

There is also a closer relationship between isometries of the field space and symmetries of the field theory. We can see this by considering the Killing vectors of (5.76). These are the solutions of Killing's equation

$$\nabla_I\xi_J + \nabla_J\xi_I = 0. \tag{5.84}$$

As shown in Sect. 2.2.1 the solutions of this equation are in one-to-one correspondence with the Noether symmetries of the theory (5.77). Let us see how these solutions compare to the Noether symmetries of (5.73).

We can write out (5.84) explicitly to expose the dependence on χ and we find

$$\begin{aligned} \nabla_A \xi_B + \nabla_B \xi_A &= 0, \\ k_{AB} \partial_\chi \xi^B + \frac{1}{V} \partial_A \xi^\chi &= 0, \\ \frac{2}{V} \partial_\chi \xi^\chi - \frac{1}{V^2} \xi^A \partial_A V &= 0, \end{aligned} \tag{5.85}$$

where ∇_A is the covariant derivative for the metric k_{AB}. In order to be a symmetry of the original theory (5.73) we look for solutions that have no dependence on the fictitious field χ so that

$$\xi^\chi = \partial_\chi \xi^A = 0. \tag{5.86}$$

We find that, with such a restriction, (5.85) reduces to

$$\nabla_A \xi_B + \nabla_B \xi_A = 0, \qquad \xi^A \partial_A V = 0. \tag{5.87}$$

These are precisely the conditions for ξ^A to be a Noether symmetry of (5.73). Thus, we have recovered the one-to-one relationship between Noether symmetries of the field theory and Killing vectors of the extended field space and there is no need for such vectors to separately leave the potential invariant.

5.5 Higher-Dimensional Field Theories

In the previous section we focused on theories with no spatial dependence. We should note that, despite their simplicity, such theories do have physical applications. The most notable example is in the theory of inflation [26–28], which is described, at leading order, by a homogeneous scalar field theory. We shall discuss inflation further in Chap. 6.

However, homogeneous field theories are obviously a severely restricted set of all field theories of interest so in this section we extend the Eisenhart lift to theories in an arbitrary number of spacetime dimensions D.

We continue to focus on a theory with N scalar fields ϕ^A and take as the action

$$S = \int d^D x \sqrt{-g} \left[\frac{1}{2} g^{\mu\nu} k_{AB}(\phi) \partial_\mu \phi^A \partial_\nu \phi^B - V(\phi) \right], \tag{5.88}$$

where $g_{\mu\nu}$ with inverse $g^{\mu\nu}$ and determinant g is the metric of spacetime, which we take to be a fixed background in this section. The equations of motion for this theory can be found by varying the action (5.88) with respect to the fields ϕ^A, which gives

$$\Box \phi^A + \Gamma^A_{BC} \partial_\mu \phi^B \partial^\mu \phi^C = -k^{AB} \partial_B V. \tag{5.89}$$

One of the key elements of the Eisenhart lift was the existence of the conserved quantity A, which the equations of motion for the fictitious field χ forced to be a constant. We wish to retain this feature when applying the Eisenhart lift to non-homogeneous fields. However, when the fields depend on D dimensions, we need D separate differential equations to force this quantity to be constant. We will, therefore, need to introduce D new degrees of freedom to provide these D conditions as their equations of motion.

We can achieve this by introducing a fictitious vector field B^μ instead of a scalar. We take the action for the extended theory to be

$$S = \int d^D x \sqrt{-g} \left[\frac{1}{2} g^{\mu\nu} k_{AB}(\phi) \partial_\mu \phi^A \partial_\nu \phi^B + \frac{1}{2} \frac{M^4}{V(\phi)} \nabla_\mu B^\mu \nabla_\nu B^\nu \right], \tag{5.90}$$

where M is, again, introduced to keep the dimensions consistent. Notice that this action is purely kinetic, and contains no potential term as expected from the Eisenhart lift.

The equations of motion for the extended theory are then

$$\Box \phi^A + \Gamma^A_{BC} \partial_\mu \phi^B \partial^\mu \phi^C = -\frac{1}{2} \left(\frac{M^2 \nabla_\mu B^\mu}{V(\phi)} \right)^2 k^{AB} \partial_B V, \tag{5.91}$$

$$\partial_\mu \left(\frac{M^2 \nabla_\nu B^\nu}{V(\phi)} \right) = 0. \tag{5.92}$$

We see that the equations of motion (5.92) that arise from the variation with respect to B^μ give exactly the D conditions we need to set

$$A = \frac{M^2 \nabla_\nu B^\nu}{V(\phi)} \tag{5.93}$$

to a constant. Plugging this condition into (5.91) gives us

$$\Box \phi^A + \Gamma^A_{BC} \partial_\mu \phi^B \partial^\mu \phi^C = -\frac{A^2}{2} k^{AB} \partial_B V, \tag{5.94}$$

which we see reduces to (5.89) provided $A^2 = 2$.

The need for the condition $A^2 = 2$ is again associated with a geometric freedom of the extended theory. However, in order to leave (5.88) invariant we must scale not only the time coordinate, but also all the spatial coordinates by the same amount. The correct transformation is therefore

$$x^\mu \to \tilde{x}^\mu = c x^\mu, \tag{5.95}$$

where c is a constant. By performing such a transformation and choosing an appropriate value for c, the condition $A^2 = 2$ can always be satisfied.

What is the geometric interpretation of the extended theory? We can construct an $N + D$ dimensional field space manifold for this theory with coordinates $\Phi^I = \{\phi^A, B^\mu\}$. Using the definition (2.41) we can equip this manifold with a metric, which we find to be

$$G_{IJ} = \begin{pmatrix} k_{AB} & 0 \\ 0 & \frac{1}{4}\frac{M^4}{V(\phi)} g_{\mu\nu} \end{pmatrix}. \tag{5.96}$$

However, because the Lagrangian for this theory is not of the form

$$L \neq \frac{1}{2} g^{\mu\nu} G_{IJ} \nabla_\mu \Phi^I \nabla_\nu \phi^J, \tag{5.97}$$

as it is for pure scalar theories, the geometric connection will not be as strong. In particular, the equations of motion for (5.90) will not reproduce the geodesic equations of (5.96) and Noether symmetries will not be in one-to-one correspondence with Killing vectors of the field space.

Let us instead see what connections we can draw by treating the field B^μ as a collection of D scalar fields. We can do this in a way consistent with Lorentz invariance by taking a set of D spacetime vectors e^μ_m (where $m \in \{1, \ldots, D\}$) and defining

$$B^\mu = e^\mu_m B^m. \tag{5.98}$$

In this way, all the vector properties of B^μ are absorbed by e^μ_m and thus B^m is a Lorentz scalar. To simplify the algebra, we shall choose vectors for which

$$\nabla_\nu e^\mu_m = 0 \tag{5.99}$$

for all μ, ν and m. Note that, in general, these vectors are different from the vielbeins of $g_{\mu\nu}$.

Thus, the action (5.88) becomes

$$S = \int d^D x \sqrt{-g} \left[\frac{1}{2} g^{\mu\nu} k_{AB}(\phi) \partial_\mu \phi^A \partial_\nu \phi^B + \frac{1}{2} \frac{M^4 e^\mu_m e^\nu_n}{V(\phi)} \partial_\mu B^m \partial_\nu B^n \right]. \tag{5.100}$$

By taking coordinates $\Phi^I = \{\phi^A, B^m\}$ and defining

$$H^{\mu\nu}_{IJ} = \begin{pmatrix} g^{\mu\nu} k_{AB} & 0 \\ 0 & \dfrac{M^4}{V} e^\mu_m e^\nu_n \end{pmatrix}, \tag{5.101}$$

we can rewrite this action as

$$S = \frac{1}{2} \int d^D x \sqrt{-g} H^{\mu\nu}_{IJ}(\Phi) \partial_\mu \Phi^I \partial_\nu \Phi^J \tag{5.102}$$

We note that $H^{\mu\nu}_{IJ} \neq g^{\mu\nu} G_{IJ}$, and hence (5.102) is not of the same form as (5.88).

Let us investigate the symmetries of (5.102) to see how they compare to the Killing equations of (5.96). We consider an infinitesimal transformation

$$\Phi^I \to \widetilde{\Phi}^I = \Phi^I + \xi^I, \tag{5.103}$$

which causes the action to change to

$$S \to \widetilde{S} = S + \delta S, \tag{5.104}$$

where

$$\delta S = \int d^D x \sqrt{-g} \left(H^{\mu\nu}_{IJ}\, \partial_K \xi^I + \frac{1}{2} \partial_I H^{\mu\nu}_{JK}\, \xi^I \right) \partial_\mu \Phi^J \partial_\nu \Phi^K . \tag{5.105}$$

For (5.103) to be a Noether symmetry of the theory, we require that $\delta S = 0$ for all field configurations. We can, therefore, compare the coefficients of $\partial_\mu \Phi^I \partial_\nu \Phi^J$ in (5.105) to get the following set of equations for ξ^I:

$$H^{\mu\nu}_{IJ}\, \partial_K \xi^I \; + \; \Gamma^{\mu\nu}_{KJI}\, \xi^I \; + \; H^{\mu\nu}_{IK}\, \partial_J \xi^I \; + \; \Gamma^{\mu\nu}_{JKI}\, \xi^I \; = \; 0 \, , \tag{5.106}$$

where

$$\Gamma^{\mu\nu}_{IJK} \; \equiv \; \frac{1}{2} \left(\partial_K H^{\mu\nu}_{IJ} \; + \; \partial_J H^{\mu\nu}_{IK} \; - \; \partial_I H^{\mu\nu}_{JK} \right) . \tag{5.107}$$

We notice that, if we treat $H^{\mu\nu}_{IJ}$ as $D(D+1)/2$ separate metrics, one for each symmetric combination of $\mu\nu$, then (5.106) tells us that a symmetry must be a Killing vector for each one. In particular, by contracting (5.106) with $g_{\mu\nu}$ we see that ξ^I must be a Killing vector for the metric

$$G_{IJ} = \frac{1}{D} g_{\mu\nu} H^{\mu\nu}_{IJ} \tag{5.108}$$

defined in (5.96).

However, this implication is only one way. While all Noether symmetries of the field theory must be Killing vectors of the field space, not all Killing vectors of the field space will necessarily be Noether symmetries of the field theory. Noether symmetries must additionally satisfy the Killing equations for all the other combinations of $H^{\mu\nu}_{IJ}$.

Finally, let us consider the equations of motion for the lifted theory. By varying (5.102) with respect to the field Φ^I, we find

$$H^{\mu\nu}_{IJ} \nabla_\mu \nabla_\nu \phi^J \; + \; \Gamma^{\mu\nu}_{IJK}\, \partial_\mu \phi^J \partial_\nu \phi^K \; = \; 0 \, . \tag{5.109}$$

When the spacetime dimension reduces to $D = 1$, we see that (5.109) reduces to the geodesic equation on the field space, as shown in the previous section. However, in general there is no such relation and, thus, we cannot interpret this system as simply a free object on the field space when the spacetime dimension is greater than one.

5.6 Quantum Field Theory

Let us now consider what happens at the quantum level. As before, it will be helpful to first reformulate the above lift in the Hamiltonian formalism. Because the Hamiltonian formalism explicitly breaks the spacetime covariance we shall consider spacetime to be flat with $g_{\mu\nu} = \eta_{\mu\nu}$ to avoid over-complicating the algebra. The action for the lifted system is, therefore,

$$S = \int d^D x \left[\frac{1}{2} k_{AB} \partial_\mu \phi^A \partial^\mu \phi_B + \frac{1}{2} \frac{M^4}{V(\phi)} \partial_\mu B^\mu \partial_\nu B^\nu \right]. \tag{5.110}$$

We begin by defining the conjugate momenta

$$\Pi_I = \frac{\delta S}{\delta \dot{\Phi}^I}, \tag{5.111}$$

with $\Phi^I = \{\phi^A, B^\mu\}$ as before. We find that

$$\Pi_I \equiv \begin{pmatrix} \pi_A \\ \pi_0 \\ \pi_i \end{pmatrix} = \begin{pmatrix} k_{AB} \dot{\phi}^B \\ \frac{M^4}{V(\phi)} \partial_\mu B^\mu \\ 0 \end{pmatrix}. \tag{5.112}$$

Here π_A is the momentum conjugate to ϕ_A, and π_0 and π_i are, respectively, the momenta conjugate to the temporal and spatial components of $\boldsymbol{B}$, which are denoted B^0 and B^i.

We notice that the spatial components of the fictitious field B^i have no conjugate momenta. This implies that they are, in fact, auxiliary fields. However, in order to derive the full set of equations of motion, we must refrain from integrating them out too early. We will therefore keep π_i in what follows, only setting it to zero at the end of our calculations.

With the canonical momenta defined, we can define the Hamiltonian

$$H = \int d^D x \left[\Pi_I \dot{\Phi}^I - \mathcal{L} \right]. \tag{5.113}$$

For the system with Lagrangian (5.110) this is

$$H = \int d^{D-1} x \left[\frac{1}{2} \left(k^{AB} \pi_A \pi_B + k_{AB} \partial_i \phi^A \partial_i \phi^B \right) + \frac{1}{2} \frac{V(\phi)}{M^4} \pi_0^2 + \pi_i \dot{B}^i - \pi_0 \partial_i B^i \right], \tag{5.114}$$

where the integral is over the $D-1$ spatial coordinates only.

Notice the appearance of $\dot{B}^i$. While Hamiltonians usually do not have time derivatives appearing explicitly, in our case we cannot write $\dot{B}^i$ in terms of any conjugate momentum and so it must remain.

We can now derive the equations of motion using Hamilton's equations

$$\dot{\Phi}^I = \frac{\delta H}{\delta \Pi_I}, \qquad \dot{\Pi}_I = -\frac{\delta H}{\delta \Phi^I}. \tag{5.115}$$

Applied to (5.114), this gives us

$$\dot{\phi}^A = k^{AB}\pi_B, \qquad \dot{\pi}_A = -\tfrac{1}{2}\partial_A k^{BC}\pi_B\pi_C - \tfrac{\pi_0^2}{2M^4}\partial_A V + k_{AB}\nabla^2\phi^B + k_{AD}\Gamma^D_{BC}\partial_i\phi^B\partial_i\phi^C, \tag{5.116}$$

$$\dot{B}^0 = \frac{V(\phi)}{M^4}\pi_0 - \partial_i B^i, \qquad \dot{\pi}_0 = 0, \tag{5.117}$$

$$\dot{B}^i = \dot{B}^i, \qquad \dot{\pi}_i = \dot{\pi}_i - \partial_i\pi_0. \tag{5.118}$$

We can put these equations into a more familiar form. First, by combining the two equations in (5.116), we arrive at

$$\Box\phi^A + \Gamma^A_{BC}\partial_\mu\phi^B\partial_\mu\phi^C = -\frac{\pi_0^2}{2M^4}k^{AB}\partial_B V, \tag{5.119}$$

where Γ^A_{BC} is defined in (5.75). Second, we can rewrite the left hand equation of (5.117) as

$$\pi_0 = \frac{M^4\partial_\mu B^\mu}{V(\phi)}. \tag{5.120}$$

Finally, we can combine the right hand equations of (5.117) and (5.118) to obtain

$$\partial_\mu\pi_0 = 0. \tag{5.121}$$

We immediately see from (5.121) that π_0 is a constant. In fact, by considering (5.120), we see that

$$\pi_0 = AM^2, \tag{5.122}$$

where A is defined in (5.93).
Furthermore, we see that when

$$\pi_0^2 = 2M^4 \tag{5.123}$$

we recover from (5.119) precisely the equations of motion of the original system with a potential. Therefore, as expected, the Eisenhart lift for field theories works just as well in the Hamiltonian formalism as it does in the Lagrangian formalism.

Let us now quantise this system in the canonical way as before. We shall promote the fields $\Phi^I(x)$ and momenta $\Pi^I(x)$ to operators $\widehat{\Phi}^I(x)$ and $\widehat{\Pi}^I(x)$, respectively. We shall infer the commutators of these operators from the Poisson brackets of their associated classical quantities as normal. This gives us

$$\left[\widehat{\phi}^A(\boldsymbol{x}), \widehat{\pi}_B(\boldsymbol{y})\right] = i\delta^A_B\delta^{(D-1)}(\boldsymbol{x}-\boldsymbol{y}), \tag{5.124}$$

$$\left[\widehat{B}^\mu(\boldsymbol{x}), \widehat{\pi}_\nu(\boldsymbol{y})\right] = i\delta^\mu_\nu\delta^{(D-1)}(\boldsymbol{x}-\boldsymbol{y}), \tag{5.125}$$

with all other commutators vanishing. In particular, all operators related to the original fields ϕ^A and their momenta π_A commute with all operators related to the fictitious fields and momenta B^μ and π_μ. This is important as it means that observations of the original fields will not affect the fictitious fields and vice versa.

Solving the full quantum dynamics of this system is incredibly challenging in general. Therefore, in the interest of clarity, we shall specialise to the case of a single canonical scalar field ϕ with potential

$$V(\phi) = \frac{1}{2}m^2\phi^2 \tag{5.126}$$

and canonical kinetic term

$$k_{AB} = 1. \tag{5.127}$$

Although simple, this theory still displays many interesting properties and can be used as a base point from which to explore more complicated systems through perturbation theory [29–31]. Since there is only one original field, we will drop the indices on $\phi_A = \phi$ and $\pi_A = \pi$ from now on to simplify notation.

The Hamiltonian operator for this theory is

$$\widehat{H} = \int d^{D-1}x \left[\frac{1}{2}\hat{\pi}^2 - \partial_i\hat{\phi}\partial_i\hat{\phi} + \frac{1}{2}m^2\hat{\phi}^2\right]. \tag{5.128}$$

This can be analysed more easily in terms of the fourier components

$$\hat{\phi}_{\boldsymbol{k}} = \int d^{D-1}x\, e^{i\boldsymbol{k}\cdot\boldsymbol{x}}\hat{\phi}(\boldsymbol{x}), \tag{5.129}$$

$$\hat{\pi}_{\boldsymbol{k}} = \int d^{D-1}x\, e^{i\boldsymbol{k}\cdot\boldsymbol{x}}\hat{\pi}(\boldsymbol{x}). \tag{5.130}$$

Here we have chosen to fourier transform only the spatial components and so $\boldsymbol{k}$ is a $D-1$ dimensional spatial vector. Performing this transformation, we can rewrite the Hamiltonian as

$$\widehat{H} = \int \frac{d^{D-1}k}{(2\pi)^{(D-1)}} \left[\frac{1}{2}\hat{\pi}_{\boldsymbol{k}}^2 + \frac{1}{2}(\boldsymbol{k}^2 + m^2)\hat{\phi}_{\boldsymbol{k}}^2\right]. \tag{5.131}$$

Comparing this expression to (5.54) we see that this is simply an infinite collection of quantum harmonic oscillators with unit mass and frequency[3]

$$\omega_{\boldsymbol{k}}^2 = \boldsymbol{k}^2 + m^2. \tag{5.132}$$

[3] This is a well-known interpretation of free quantum field theories. See, for example [2], for further details.

We should, therefore, be able to analyse the system in terms of ladder operators as we did before. Indeed, the relevant operators are given by

$$\hat{a}_k = \sqrt{\frac{\omega_k}{2}}\hat{\phi}_k + \frac{i}{\sqrt{2\omega_k}}\hat{\pi}_k, \tag{5.133}$$

$$\hat{a}_k^\dagger = \sqrt{\frac{\omega_k}{2}}\hat{\phi}_k - \frac{i}{\sqrt{2\omega_k}}\hat{\pi}_k. \tag{5.134}$$

A common interpretation of these operators is that $\hat{a}_k^\dagger$ creates particle with momentum $\boldsymbol{k}$ and energy ω_k and $\hat{a}_k$ annihilates that same particle. They are, therefore, often known as the creation and annihilation operators, respectively.

Armed with these operators, we can rewrite the Hamiltonian as

$$\widehat{H} = \int \frac{d^{D-1}k}{(2\pi)^{(D-1)}}\left[\omega_k\left(\hat{a}_k^\dagger\hat{a}_k + \frac{1}{2}[\hat{a}_k, \hat{a}_k^\dagger]\right)\right]. \tag{5.135}$$

The commutator in the second term gives rise to a divergence proportional to $\delta^{(D-1)}(0)$. However, this zero point energy cannot be detected experimentally and so we shall ignore it, as is commonly done in the literature [32].[4]

Let us now introduce the fictitious quantum vector field $\widehat{B}^\mu$ and apply the Eisenhart lift to this system. The resulting Hamiltonian is, therefore,

$$\widehat{H} = \int d^{D-1}x\left[\frac{1}{2}\hat{\pi}^2 - \partial_i\hat{\phi}\partial_i\hat{\phi} + \frac{m^2}{2M^4}\hat{\phi}^2\hat{\pi}_0^2 + \hat{\pi}_i\dot{\hat{B}}^i - \hat{\pi}_0\partial_i\hat{B}^i\right]. \tag{5.136}$$

From our study of the classical system, we know that π_0 is conserved in both space and time. To understand whether this persists at the quantum level, we need to consider the commutator of $\hat{\pi}_0$ with the momentum operator

$$\widehat{P}_\mu = \int d^{D-1}x\widehat{T}_{0\mu}. \tag{5.137}$$

Here

$$\widehat{T}_{\mu\nu} = \frac{\partial\widehat{\mathcal{L}}}{\partial(\partial^\mu\widehat{\Phi}^I)}\partial_\nu\widehat{\Phi}^I - \eta_{\mu\nu}\widehat{\mathcal{L}} \tag{5.138}$$

is the energy momentum tensor.

For the system with Hamiltonian (5.136) we have

[4] Note that the nature of this zero point energy is still very poorly understood. It predicts an enormous value for the cosmological constant, which is in stark contradiction with observations. This issue is known as the cosmological constant problem and is a highly active area of research [33, 35, 36].

$$\widehat{P}_0 = \widehat{H} \tag{5.139}$$

$$\widehat{P}_i = \int d^{D-1}x \left[\hat{\pi}\partial_i\hat{\phi} + \hat{\pi}_0\partial_i\widehat{B}^0\right]. \tag{5.140}$$

We can immediately see that

$$[\hat{\pi}_0(\boldsymbol{x}), \widehat{P}_0] = 0, \tag{5.141}$$

since the Hamiltonian (5.136) has no dependence on $\widehat{B}^0$. The commutator with the spatial momentum gives

$$[\hat{\pi}_0(\boldsymbol{x}), \widehat{P}_i] = -i\int d^{D-1}y\left[\hat{\pi}_0(\boldsymbol{y})\partial_i\delta^{(D-1)}(\boldsymbol{y}-\boldsymbol{x})\right] \tag{5.142}$$

$$= i\partial_i\hat{\pi}_0(\boldsymbol{x}). \tag{5.143}$$

When applied to any physical on-shell state, the equation of motion (5.121) tells us that this will give zero.

Thus, the commutator

$$[\widehat{P}_\mu, \hat{\pi}_0] = 0 \tag{5.144}$$

vanishes when acting on any physical state for all values of the spacetime index μ. This means that π_0 is a good quantum number, conserved in both time and space. We can, therefore, drop any spacetime dependence of the operator $\hat{\pi}_0$.

The final term in (5.136) is, therefore, a total derivative and can be removed. In addition, we can remove the penultimate term, since we know the operator $\hat{\pi}_i$ will yield zero when acting on any physical state. Hence, we can rewrite the Hamiltonian (5.136) as

$$\widehat{H} = \int \frac{d^{D-1}k}{(2\pi)^{(D-1)}}\left[\frac{1}{2}\hat{\pi}_{\boldsymbol{k}}^2 + \frac{1}{2}\left(\boldsymbol{k}^2 + \frac{m^2}{M^4}\hat{\pi}_0^2\right)\hat{\phi}_{\boldsymbol{k}}^2\right], \tag{5.145}$$

where we have Fourier transformed $\hat{\phi}$ and $\hat{\pi}$ using (5.129) and (5.130) as before.

We see that (5.145) is very similar to (5.131) and we should, therefore, be able to analyse it in terms of creation and annihilation operators. In fact, by taking inspiration from (5.66) and (5.67) we see that these are

$$\hat{\mathfrak{a}}_{\boldsymbol{k}} \equiv \frac{\sqrt{\omega_{\boldsymbol{k}}}}{2M^2}\hat{\pi}_0\hat{\phi}_{\boldsymbol{k}} + \frac{i}{\sqrt{2\omega_{\boldsymbol{k}}}}\hat{\pi}_{\boldsymbol{k}}, \tag{5.146}$$

$$\hat{\mathfrak{a}}_{\boldsymbol{k}}^\dagger \equiv \frac{\sqrt{\omega_{\boldsymbol{k}}}}{2M^2}\hat{\pi}_0\hat{\phi}_{\boldsymbol{k}} - \frac{i}{\sqrt{2\omega_{\boldsymbol{k}}}}\hat{\pi}_{\boldsymbol{k}}. \tag{5.147}$$

This allows us to rewrite the Hamiltonian (5.145) as

$$\widehat{H} = \int \frac{d^{D-1}k}{(2\pi)^{(D-1)}}\left[\omega_{\boldsymbol{k}}\left(\hat{\mathfrak{a}}_{\boldsymbol{k}}^\dagger\hat{\mathfrak{a}}_{\boldsymbol{k}} + \frac{1}{2}[\hat{\mathfrak{a}}_{\boldsymbol{k}}^\dagger, \hat{\mathfrak{a}}_{\boldsymbol{k}}]\right)\right]. \tag{5.148}$$

The second term is again divergent but, as before, this will not affect physical observables.

Therefore, the lifted system can still be described by a set of particles with $\hat{\mathfrak{a}}_{\boldsymbol{k}}^{\dagger}$ and $\hat{\mathfrak{a}}_{\boldsymbol{k}}$ respectively creating and destroying a particle with momentum $\boldsymbol{k}$. The energy of this particle can be calculated from

$$\widehat{H}\,\hat{\mathfrak{a}}_{\boldsymbol{k}}^{\dagger}\,|0\rangle = [\widehat{H}, \hat{\mathfrak{a}}_{\boldsymbol{k}}^{\dagger}]\,|0\rangle = \frac{\omega_{\boldsymbol{k}}}{\sqrt{2M^2}}\hat{\pi}_0\hat{\mathfrak{a}}_{\boldsymbol{k}}^{\dagger}\,|0\rangle\,. \tag{5.149}$$

Here we have defined $|0\rangle$ to be the vacuum state and used the fact that $\widehat{H}\,|0\rangle = 0$. We have previously discovered that any solution to the TISE will be an eigenstate of the operator $\hat{\pi}_0$. If we call the corresponding eigenvalue π_0 then we find the energy of the particle is

$$E_{\boldsymbol{k},\pi_0} = \frac{\pi_0}{\sqrt{2M^2}}\omega_{\boldsymbol{k}}. \tag{5.150}$$

As we have seen, the value of π_0 is constant in time and space and is, therefore, a fixed value for the entire Universe. If that value satisfies the Eisenhart condition (5.123) then the spectrum of the lifted theory is identical to that of the original theory. If not, then we see that all energy levels must be multiplied by $\pi_0/\sqrt{2M^2}$. Since this is a universal rescaling, it will be equivalent to a change of energy units which, as argued in Chap. 1 is unobservable. Therefore, we see that the lifted system is fully equivalent to the original system at both the classical and quantum levels.

5.7 Further Generalisations of the Eisenhart Lift

The techniques of the Eisenhart lift generalise readily to theories with fermion, vector or tensor degrees of freedom. We see that the spacetime properties of the fields never entered into the derivation of (5.90) and, thus, a potential can be transformed into a kinetic term for *any* QFT, regardless of its field content. In fact, the Eisenhart lift doesn't even need to be applied to the potential term as we will show below.

Let us consider a field theory with a set of fields Φ^{α}, collectively denoted $\boldsymbol{\Phi}$. We will make no assumptions about the spacetime properties of these fields, nor whether they are commutative or anticommutative. We take the action of this theory to be

$$S = \int d^D x \sqrt{-\bar{g}}\left[\mathcal{L}_1(\boldsymbol{\Phi}, \partial_\mu \boldsymbol{\Phi}) + \mathcal{L}_2(\boldsymbol{\Phi}, \partial_\mu \boldsymbol{\Phi})\right], \tag{5.151}$$

where $\mathcal{L}_1$ and $\mathcal{L}_2$ are arbitrary functions of the fields and their derivatives and $\bar{g}_{\mu\nu}$ with determinant $\bar{g}$ is the spacetime metric.

We now repeat the construction in Sect. 5.5 by introducing a new vector field B^{μ} to our theory. We shall replace the term $\mathcal{L}_2$ in the action (5.151) with a kinetic term for B^{μ}. We, thus, define

$$S = \int d^D x \sqrt{-\bar{g}} \left[\mathcal{L}_1 - \frac{1}{2} \frac{M^4}{\mathcal{L}_2} \overline{\nabla}_\mu B^\mu \overline{\nabla}_\nu B^\nu \right], \tag{5.152}$$

where M is an arbitrary mass scale as before and $\overline{\nabla}$ is the covariant derivative for the metric $\bar{g}_{\mu\nu}$.

Note that in order to do this we have had to divide by $\mathcal{L}_2$. One may worry if such an operation is well defined when the theory contains fermions, vector fields or tensor fields, since such fields do not always have a well defined inverse. However, since $\mathcal{L}_2$ is part of the Lagrangian, it must be a commuting Lorentz scalar regardless of the properties of the fields it depends on. Therefore, dividing by $\mathcal{L}_2$ is a well-defined operation.[5]

We can vary (5.152) with respect to B^μ to obtain the following equation of motion:

$$\partial_\mu \left(\frac{M^2 \overline{\nabla}_\nu B^\nu}{\mathcal{L}_2(\mathbf{\Phi}, \partial_\mu \mathbf{\Phi})} \right) = 0. \tag{5.153}$$

This is very similar to (5.92) and yields a similar solution

$$A \equiv \frac{M^2 \overline{\nabla}_\mu B^\mu}{\mathcal{L}_2(\mathbf{\Phi}, \partial_\mu \mathbf{\Phi})} = \text{constant}. \tag{5.154}$$

Note that regardless of the nature of the field content, A is a commuting Lorentz scalar.

Similarly, we can vary (5.152) with respect to Φ^α to obtain the other equations of motion:

$$\overline{\nabla}_\mu \left(\frac{\overrightarrow{\partial}}{\partial(\partial_\mu \Phi^\alpha)} \mathcal{L}_1 \right) - \frac{\overrightarrow{\partial}}{\partial \Phi^\alpha} \mathcal{L}_1 + \overline{\nabla}_\mu \left(\frac{A^2}{2} \frac{\overrightarrow{\partial}}{\partial(\partial_\mu \Phi^\alpha)} \mathcal{L}_2 \right) - \frac{A^2}{2} \frac{\overrightarrow{\partial}}{\partial \Phi^\alpha} \mathcal{L}_2 = 0. \tag{5.155}$$

Here, we have expressed the equations of motion in terms of the left derivative, however, a similar expression can also be obtained in terms of the right derivative. There will only be a difference between left and right derivatives if the theory contains fermions.

As we have seen, (5.154) tells us that A is a constant, commuting Lorentz scalar. If we choose that constant to satisfy $A^2 = 2$ we see that (5.155) reduces to

$$\overline{\nabla}_\mu \left(\frac{\overrightarrow{\partial}}{\partial(\partial_\mu \Phi^\alpha)} (\mathcal{L}_1 + \mathcal{L}_2) \right) - \frac{\overrightarrow{\partial}}{\partial \Phi^\alpha} (\mathcal{L}_1 + \mathcal{L}_2) = 0, \tag{5.156}$$

which are precisely the equations of motion for (5.151).

[5] There is one exception to this, which is if $\mathcal{L}_2$ is nilpotent. However, such cases are rare and so we will not consider them here.

We have, therefore, shown that it is possible to replace *any* term, or even a collection of terms, in the action of a QFT with the kinetic term of a vector field.

5.7.1 Example: Scalar Fermion Theory

As an example we consider a scalar fermion theory as discussed in Chap. 4. We consider a theory with N real scalar fields ϕ^A and M Dirac fermions ψ_a^X and denote them collectively by $\Phi^\alpha = \{\phi^A, \psi_a^X, \overline{\psi}_{\dot{a}}^X\}$ as before. We take the action to be (4.35), repeated here for convenience:

$$S = \int d^D x \sqrt{-g} \left[\frac{1}{2} g^{\mu\nu} \partial_\mu \Phi^\alpha \; {}_\alpha k_\beta(\mathbf{\Phi}) \; \partial_\nu \Phi^\beta + \frac{1}{2} \zeta_\alpha^\mu(\mathbf{\Phi}) \; \partial_\mu \Phi^\alpha - U(\mathbf{\Phi}) \right]. \tag{5.157}$$

Varying (5.157) with respect to Φ^α tells us that the equations of motion for this theory are

$$\overline{\nabla}_\mu \left({}_\alpha k_\beta \partial^\mu \Phi^\beta \right) + (-1)^\alpha \frac{1}{2} \overline{\nabla}_\mu \zeta_\alpha^\mu + \overrightarrow{\partial}_\alpha U = 0. \tag{5.158}$$

We wish to eliminate the potential term $U(\mathbf{\Phi})$ so we follow the above procedure with

$$\mathcal{L}_1 = \frac{1}{2} g^{\mu\nu} \partial_\mu \Phi^\alpha \; {}_\alpha k_\beta(\mathbf{\Phi}) \; \partial_\nu \Phi^\beta + \frac{1}{2} \zeta_\alpha^\mu(\mathbf{\Phi}) \partial_\mu \Phi^\beta, \qquad \mathcal{L}_2 = U(\mathbf{\Phi}). \tag{5.159}$$

We add a vector field B^μ to the theory and define the purely kinetic action

$$S = \int d^D x \sqrt{-g} \left[\frac{1}{2} g^{\mu\nu} \partial_\mu \Phi^\alpha \; {}_\alpha k_\beta(\mathbf{\Phi}) \; \partial_\nu \Phi^\beta + \frac{1}{2} \zeta_\alpha^\mu(\mathbf{\Phi}) \; \partial_\mu \Phi^\alpha + \frac{1}{2} \frac{1}{U(\mathbf{\Phi})} \nabla_\mu B^\mu \nabla_\nu B^\nu \right]. \tag{5.160}$$

Note that, although $U(\mathbf{\Phi})$ can depend on the fermion fields ψ^X it is still a commuting number and a Lorentz scalar and, thus, can only depend on fermion bilinears. This means its inverse can be well defined.[6]

We can derive the equations of motion for (5.160) and we find

$$\nabla_\mu \left(g^{\mu\nu} {}_\alpha k_\beta \partial_\nu \Phi^\beta \right) + (-1)^\alpha \frac{1}{2} \nabla_\mu \zeta_\alpha^\mu + \frac{(\nabla_\mu B^\mu)^2}{2U^2} \overrightarrow{\partial}_\alpha U = 0, \tag{5.161}$$

$$\partial_\mu \left(\frac{\nabla_\nu B^\nu}{U(\mathbf{\Phi})} \right) = 0. \tag{5.162}$$

---

[6] As noted before, there is an exception to this if $U(\mathbf{\Phi})$ is a nilpotent number, for example if it contains only mass terms for the fermions and no scalar potential. We will not consider this possibility here.

Equation (5.162) implies

$$A = \frac{\nabla_\mu B^\mu}{U(\mathbf{\Phi})} = \text{constant}. \tag{5.163}$$

If we choose this constant to satisfy $A^2 = 2$ then (5.161) becomes identical to (5.158). Thus, we are able to construct a purely kinetic Lagrangian that yields the same dynamics as any theory of fermion and scalar fields.

References

1. Einstein A (1915) The field equations of gravitation. Sitzungsber Preuss Akad Wiss Berlin (Math Phys) 1915:844–847 (cit. on pp. 19, 39, 138, 174)
2. Peskin ME, Schroeder DV (1995) An introduction to quantum field theory. Front Phys. Avalon Publishing (cit. on pp. 25, 59, 162)
3. Kaluza T (1921) Zum Unitätsproblem der Physik. Sitzungsber Preuss Akad Wiss Berlin (Math Phys) 1921:966–972. arXiv: 1803. 08616 [physics.hist-ph] (cit. on pp. 83, 139)
4. Klein O (1926) Quantum theory and five-dimensional theory of relativity. (In German and English). Z Phys 37:895–906. In: Taylor JC (ed) (cit. on pp. 83, 139)
5. Eisenhart LP (1928) Dynamical trajectories and geodesics. Ann Math 30(1/4):591–606 (cit. on pp. 135, 136)
6. Duval C, Burdet G, Kunzle HP, Perrin M (1985) Bargmann structures and newton-cartan theory. Phys Rev D 31:1841–1853 (cit. on p. 135)
7. Minguzzi E (2007) Eisenhart's theorem and the causal simplicity of Eisenhart's spacetime. Class Quant Grav 24:2781–2808. arXiv: grqc/ 0612014 (cit. on p. 135)
8. Cariglia M, Gibbons G (2014) Generalised Eisenhart lift of the Toda chain. J Math Phys 55:022701. arXiv: 1312.2019 [math-ph] (cit. on p. 135)
9. Cariglia M, Duval C, Gibbons GW, Horvathy PA (2016) Eisenhart lifts and symmetries of time-dependent systems. Ann Phys 373:631–654. arXiv: 1605.01932 [hep-th] (cit. on p. 135)
10. Cariglia M, Galajinsky A, Gibbons GW, Horvathy PA (2018) Cosmological aspects of the Eisenhart-Duval lift. Eur Phys J C 78(4):314. arXiv: 1802.03370 [gr-qc] (cit. on p. 135)
11. Cariglia M, Alves FK (2015) The Eisenhart lift: a didactical introduction of modern geometrical concepts from Hamiltonian dynamics. Eur J Phys 36(2):025018. arXiv: 1503.07802 [gr-qc] (cit. on p. 135)
12. Newton I (1687) Philosophiae naturalis principia mathematica. J Societatis Regiae ac Typis J Streater (cit. on pp. 136, 138)
13. Coriolis G-G (1835) Sur les équations du mouvement relatif des systèmes de corps. J De l'Ecole Polytech 15:144–154 (cit. on p. 138)
14. Overduin JM, Wesson PS (1997) Kaluza-Klein gravity. Phys Rept 283:303–380. arXiv: gr-qc/9805018 [gr-qc] (cit. on p. 139)
15. Hamilton WR (1834) On a general method in dynamics. Phil Trans R Soc 124:247–308 (cit. on p. 139)
16. Hamilton WR (1835) Second essay on a general method in dynamics. Phil Trans R Soc 125:95–144 (cit. on p. 139)
17. Dirac PAM (1925) The fundamental equations of quantum mechanics. Proc Roy Soc Lond A 109:642–653 (cit. on p. 140)
18. Dirac PAM (1930) The principles of quantum mechanics. Oxford University Press (cit. on pp. 140, 148)
19. Schrödinger E (1926) Quantisierung als Eigenwertproblem. Annalen Phys 384(4):361–376 (cit. on p. 141)

20. Schrödinger (1926) Quantisierung als Eigenwertproblem. Annalen Phys 384(6):489–527 (cit. on p. 141)
21. Schrödinger E (1926) Quantisierung als Eigenwertproblem. Annalen Phys 385(13):437–490 (cit. on p. 143)
22. Schrödinger E (1926) Quantisierung als Eigenwertproblem. Annalen Phys 386(18):109–139 (cit. on p. 143)
23. Griffiths DJ, Schroeter DF (2018) Introduction to quantum mechanics. Cambridge University Press (cit. on p. 148)
24. Gasiorowicz S (2003) Quantum physics. Wiley (cit. on p. 148)
25. Laplace PS (1810) Mémoire sur les intégrales définies et leur application aux probabilités, et spécialement a la recherche du milieu qu'il faut choisir entre les résultats des observations. Mémoires de l'Académie des Sci 279–347 (cit. on p. 148)
26. Guth AH (1981) The inflationary universe: a possible solution to the horizon and flatness problems. Phys Rev D23:347–356 (cit. on pp. 154, 184, 249)
27. Linde AD (1982) A new inflationary universe scenario: a possible solution of the horizon, flatness, homogeneity, isotropy and primordial monopole problems. Phys Lett 108B:389–393 (cit. on pp. 154, 184, 249)
28. Albrecht A, Steinhardt PJ (1982) Cosmology for grand unified theories with radiatively induced symmetry breaking. Phys Rev Lett 48:1220–1223 (cit. on pp. 154, 184, 249)
29. Gell-Mann M, Low F (1951) Bound states in quantum field theory. Phys Rev 84(2):350–354 (cit. on p. 161)
30. Molinari LG (2007) Another proof of Gell-Mann and Low's theorem. J Math Phys 48(5):052113 (cit. on p. 161)
31. Fetter AL, Walecka JD (2003) Quantum theory of many-particle systems. In: Dover books on physics. Dover Publications (cit. on p. 161)
32. Sciama DW (1991) The physical significance of the vacuum state of a quantum field. In: The philosophy of vacuum. Oxford University Press, pp 137–158 (cit. on p. 162)
33. Martin J (2012) Everything you always wanted to know about the cosmological constant problem (But were afraid to ask). Comptes Rendus Phys 13:566–665. arXiv:1205.3365 [astro-ph.CO] (cit. on pp. 162, 233, 249)
34. Sola J (2013) Cosmological constant and vacuum energy: old and new ideas. J Phys Conf Ser 453:012015. In: Papakostas T, Pliakis DA (eds). arXiv: 1306.1527 [gr-qc] (cit. on pp. 162, 233, 249)
35. Frusciante N, Perenon L (2020) Effective field theory of dark energy: a review. Phys Rept 857:1–63. arXiv: 1907.03150 [astro-ph.CO] (cit. on pp. 162, 233, 249)
36. Carroll SM (2001) The cosmological constant. Living Rev Rel 4:1. arXiv: astro-ph/0004075 (cit. on pp. 162, 233, 249)
37. Kieran Finn, Sotirios Karamitsos, and Apostolos Pilaftsis. "Eisenhart lift for field theories". In: Phys. Rev. D 98.1 (2018), p. 016015. arXiv: 1806.02431 [physics.class-ph] (cit. on p. 147)

Chapter 6
Cosmic Inflation

Space is big. You just won't believe how vastly, hugely, mind-bogglingly big it is. I mean, you may think it's a long way down the road to the chemist's, but that's just peanuts to space.

—Douglas Adams

One of the first applications of the geometric description of QFTs was to the theory of inflation [2, 3, 14–17]. In this chapter we review early Universe cosmology and the motivation behind the theory of inflation, including the classic fine-tuning puzzles known as the horizon, flatness and monopole problems. We will show how inflation purports to solve these puzzles before reviewing the current debate on whether inflation introduces new fine tuning issues of its own.

6.1 Hot Big Bang Cosmology

In accordance with General Relativity, the Universe is best described in terms of a spacetime manifold. The standard theory of cosmology posits that, at leading order, this manifold is completely homogeneous and isotropic. The most general metric consistent with these assumptions is the Friedmann Robertson Walker (FRW) metric [4–10]:

$$ds^2 = dt^2 - a(t)^2 \left[\frac{1}{1 - kr^2} dr^2 + r^2 d\theta^2 + r^2 \sin^2 \theta d\phi^2 \right]. \tag{6.1}$$

Here, $a(t)$ is the *scale factor* and k is a constant.

K. Finn, *Geometric Approaches to Quantum Field Theory*, Springer Theses,
https://doi.org/10.1007/978-3-030-85269-6_6

The value of k defines the spatial curvature of the Universe. We can see this by calculating Ricci scalar for a three-dimensional spatial slice $t = t_* =$ constant, which is given by

$$R_3 = \frac{6}{a(t_*)^2}k. \tag{6.2}$$

Note that by redefining the radial coordinate and the scale factor we can rescale the value of k to whatever we like as long as we do not change its sign. It is conventional to use this freedom to set $k = 0$, $k = 1$ or $k = -1$.

Similarly, since the overall normalisation of (6.1) is irrelevant, we can change the overall scale of $a(t)$ arbitrarily by rescaling the time coordinate. This freedom is conventionally used to set $a(t_0) = 1$, where t_0 refers to today.

To calculate the evolution the Universe we employ Einstein's equation [1]

$$R_{\mu\nu} - \frac{1}{2}g_{\mu\nu}R = T_{\mu\nu}. \tag{6.3}$$

Here, $R_{\mu\nu}$ and R are the Ricci tensor and scalar, respectively, and

$$T_{\mu\nu} = 2\frac{\partial \mathcal{L}_m}{\partial g^{\mu\nu}} - g_{\mu\nu}\mathcal{L}_m \tag{6.4}$$

is the stress energy tensor of the non-gravitational (matter) content of the Universe, which has Lagrangian $\mathcal{L}_m$.

We can calculate the Ricci tensor and scalar for the metric (6.1) and we find them to be [18]

$$R_{\mu\nu} = \begin{pmatrix} 3\frac{\ddot{a}}{a} & 0 & 0 & 0 \\ 0 & \frac{a\ddot{a}+2\dot{a}^2+2k}{1-kr^2} & 0 & 0 \\ 0 & 0 & r^2(a\ddot{a} + 2\dot{a}^2 + 2k) & 0 \\ 0 & 0 & 0 & r^2\sin^2\theta(a\ddot{a} + 2\dot{a}^2 + 2k) \end{pmatrix}, \tag{6.5}$$

$$R = -6\left(H^2 + \frac{\ddot{a}}{a} + \frac{k}{a^2}\right), \tag{6.6}$$

where a dot $\dot{}$ indicates differentiation with respect to time and $H \equiv \dot{a}/a$ is the Hubble parameter.

The stress-energy tensor (also known as the energy-momentum tensor) describes the spatial density of energy and momentum. Mathematically it is given by

$$\delta p_\mu = T_{\mu\nu}\delta V_3^\mu, \tag{6.7}$$

where δp_μ is the four-momentum contained in the infinitesimal 3-volume element δV_3^μ. Note that in four-dimensional spacetime, a three dimensional volume is described by a 4-vector, just as a two dimensional area is described by a 3-vector in 3D space.

The components of the stress energy tensor can be physically understood as follows:

$$T_{\mu\nu} = \left(\begin{array}{c|c} \text{energy density} & \text{energy flux} \\ \hline \text{momentum density} & \text{stress tensor} \end{array} \right) \tag{6.8}$$

where the stress tensor T_{ij} is the ith component of the force per unit area exerted across a surface with normal in direction j.

The only matter content consistent with the assumptions of homogeneity and isotropy is a homogeneous ideal fluid at rest. If such a fluid has a spatial density ρ and exerts a pressure p then its stress energy tensor will be given by

$$T_{\mu\nu} = \begin{pmatrix} \rho & 0 & 0 & 0 \\ 0 & p\frac{a^2}{1-kr^2} & 0 & 0 \\ 0 & 0 & pa^2r^2 & 0 \\ 0 & 0 & 0 & pa^2r^2\sin^2\theta \end{pmatrix}. \tag{6.9}$$

We can now derive the evolution equations of the Universe from (6.3). Starting with the 00 component we find

$$H^2 = \frac{1}{3}\rho - \frac{k}{a^2}. \tag{6.10}$$

Notice that there are no second derivatives in this equation. This equation is, therefore, not a dynamical evolution equation, but rather a constraint. For this reason it is sometimes known as the Hamiltonian constraint.

The rest of the diagonal components of (6.3) all yield the same equation

$$p = -2\frac{\ddot{a}}{a} - H^2 - \frac{k}{a^2}. \tag{6.11}$$

It is conventional to eliminate the curvature term in this equation with (6.10) and so derive the Raychaudhuri equation [19]

$$\frac{\ddot{a}}{a} = -\frac{1}{6}(\rho + 3p). \tag{6.12}$$

Collectively the Eqs. (6.10)–(6.12) are known as the Friedmann equations.

There are no off-diagonal terms in (6.3) and so the Friedmann equations constitute the full set of Einstein's equations. However, in their current state, these equations are underdetermined. We are, therefore, unable to proceed without making a further assumption about the matter content of the Universe. A common assumption to make is that the Universe is dominated by a single type of matter with constant equation of state w, so that

$$p = w\rho. \tag{6.13}$$

With this assumption, the Eqs. (6.10)–(6.12) can be solved and the dynamics of the Universe determined.

We start by differentiating (6.10) which, with the help of (6.11), gives

$$\dot{\rho} + 3H(\rho + p) = 0. \tag{6.14}$$

This is commonly known as the continuity equation. We can rewrite (6.14) with the help of (6.13) and the chain rule to give

$$\frac{d\rho}{da} = -3\frac{\rho}{a}(1 + w). \tag{6.15}$$

This equation is straightforward to solve and we find

$$\rho = \rho_0 a^{-3(1+w)}, \tag{6.16}$$

where ρ_0 is the density today.

The curvature term $-k/a^2$ in (6.10) has the form of a density (6.16). We can, therefore, interpret curvature as an ideal fluid with $\rho_0 = -3k$ and $w = -1/3$ and, hence, treat it as just another contribution to the energy density of the Universe. The two terms on the RHS of (6.10) can then be absorbed into one, giving us

$$\dot{a} = \sqrt{\frac{\rho_0}{3}a^{-\frac{1}{2}(1+3w)}}. \tag{6.17}$$

We can solve (6.17), taking into account the convention $a(t_0) = 1$, to obtain

$$a(t) = \left(\frac{t}{t_0}\right)^{\frac{2}{3(1+w)}}, \tag{6.18}$$

and therefore

$$H = \frac{2}{3(1 + w)}t^{-1} \tag{6.19}$$

for all values of the equation of state $w \neq -1$.

For the specific case of $w = -1$, we see from (6.16) that ρ, and therefore H is a constant. Such a fluid is, therefore, known as a cosmological constant, and is usually denoted by Λ. During Λ domination we see that the scale factor grows exponentially with $a \propto \exp(Ht)$.

In addition to curvature and the cosmological constant, the other common types of fluid to be considered are non-relativistic matter, which is pressureless with $w = 0$ and radiation (or equivalently relativistic matter), which has $w = 1/3$. The density of these components, therefore, dilute as $\rho_m \propto a^{-3}$ and $\rho_r \propto a^{-4}$, respectively. The evolution properties of all common forms of matter are shown in Table 6.1.

Table 6.1 Cosmic evolution of different forms of matter

	w	$\rho(a)$	$a(t)$	$H(t)$	$H(a)$
Matter	0	a^{-3}	$t^{\frac{2}{3}}$	$\frac{2}{3}t^{-1}$	$a^{-\frac{3}{2}}$
Radiation	$\frac{1}{3}$	a^{-4}	$t^{\frac{1}{2}}$	$\frac{1}{2}t^{-1}$	a^{-2}
Curvature	$-\frac{1}{3}$	a^{-2}	t	t^{-1}	a^{-1}
Λ	-1	const.	e^{Ht}	const.	const.
General	w	$a^{-3(1+w)}$	$t^{\frac{2}{3(1+w)}}$	$\frac{2}{3(1+w)}t^{-1}$	$a^{-\frac{3(1+w)}{2}}$

We can also trace the thermal history of the Universe. In the early Universe all matter was in thermal contact and, thus, formed a thermal bath at some temperature T. This bath is still seen today in the form of the Cosmic Microwave Background (CMB) [20].

We can calculate how the temperature of the CMB has evolved by considering Planck's law for the spectral distribution of black body radiation [21]

$$B(\lambda, T) = \frac{2}{5}\frac{1}{e^{\frac{1}{\lambda T}} - 1}. \tag{6.20}$$

Integrating Planck's distribution, we find that the density of radiation from the CMB is

$$\rho_r = \frac{\pi^2}{30}T^4. \tag{6.21}$$

But, from Table 6.1 we know that this density dilutes as $\rho \propto a^{-4}$ as the Universe evolves. We, therefore, see that the temperature of the Universe dilutes as

$$T \propto a^{-1}. \tag{6.22}$$

In the past when the Universe was smaller, it was, therefore, also hotter. This is the origin of the term Hot Big Bang (HBB).

Let us end this section by reviewing the consequences of the above calculations on the history of the Universe. We see from Table 6.1 that the total energy density in the Universe is given by

$$\rho = \frac{\rho_{r,0}}{a^4} + \frac{\rho_{m,0}}{a^3} + \Lambda, \tag{6.23}$$

where $\rho_{r,0}$ and $\rho_{m,0}$ are, respectively, the density of radiation and matter today and Λ is the cosmological constant. In the early Universe, when $a \ll 1$ we see that the Universe was dominated by radiation and so grew as $a \propto t^{\frac{1}{2}}$. At some point the Universe expanded sufficiently that $a > \rho_{r,0}/\rho_{m,0}$ and we entered the matter-dominated phase of the Universe with $a \propto t^{\frac{2}{3}}$. In the future, the Universe will continue to expand, diluting both the radiation and matter densities until only the cosmological constant remains, leaving the Universe to expand exponentially forever.

Today, observations tell us that we are at the transition between matter and Λ domination with an energy density that is 69% cosmological constant, 31% non-relativistic matter and a negligible amount of radiation [22].

Note that we have ignored curvature in the previous discussion. In theory there could have been a period of curvature domination between the matter era and the cosmological constant. In fact, had the curvature been sufficiently positive, it would have caused the Universe to stop expanding altogether and recollapse in a Big Crunch. However, current observations show that the curvature of our Universe is incredibly small and so has never been a significant portion of the energy density. This is actually a surprising discovery as we shall see in the next section.

6.2 The Classic Cosmological Puzzles

The HBB model of cosmology described above has proven very successful. It explains the current expansion of the Universe, first observed by Hubble [23], as the consequence Einstein's equations. It explains the origin of the Cosmic Microwave Background (CMB) radiation observed by Penzias and Wilson [20] as the radiation component of the energy budget of the Universe. It also explains, in exquisite detail, the relative abundances of the light elements through the process of Big Bang Nucleosynthesis (BBN) [24, 25].

However, there are still many shortcomings of this model, some of which are detailed below.

6.2.1 The Flatness Problem

The first puzzle relates to the observed flatness of the Universe today. The problem can be best understood by defining the density parameter

$$\Omega = \frac{\rho}{3H^2}. \tag{6.24}$$

With this definition, (6.10) can be recast in the form

$$1 - \Omega = -\frac{k}{a^2 H^2}. \tag{6.25}$$

Equation (6.25) tells us that we can infer the curvature of our Universe by measuring its density today and comparing it to the critical density $\rho_{crit} = 3H_0^2$, where H_0 is the value of the Hubble parameter today. If we measure $\rho > \rho_{crit}$ then we must have $k = +1$. If $\rho < \rho_{crit}$ then $k = -1$. The boundary case $\rho = \rho_{crit}$ corresponds to a flat universe with $k = 0$.

Measurements by the Planck satellite [22] tell us that the current density of the Universe is incredibly close to the critical density with

$$1 - \Omega_0 = -0.056^{+0.028}_{-0.018}\ (CMB), \qquad 0.0007^{+0.0019}_{-0.0019},\ (CMB + BAO), \tag{6.26}$$

where the first value comes solely from CMB measurements and the second value also includes measurements of the Baryon Acoustic Oscillations (BAO) from galaxy surveys [26]. In both cases the interval shown is the 68% confidence level.

In fact, the situation is even more problematic. We can see from (6.25) that $1 - \Omega$ evolves over time as $(aH)^{-2}$. During the matter and radiation dominated eras we see this factor scaled as $(aH)^{-2} \propto t^{\frac{2}{3}}$ and $(aH)^{-2} \propto t$, respectively, and therefore $1 - \Omega$ grew during both of these periods. This means the Universe was even flatter in the past and, in particular

$$1 - \Omega_P = \left(\frac{(aH)_0}{(aH)_P}\right)^2 (1 - \Omega_0) \sim 10^{-62} \tag{6.27}$$

at the Planck time (denoted here by the subscript P).

We, therefore, see that, in order to explain the observed flatness today, the Universe must have begun with an incredibly finely tuned initial density at the Planck scale, equal to the critical density to one part in 10^{62}. The HBB model has no explanation for this fine tuning.

6.2.2 *The Horizon Problem*

In the HBB model the Universe has a finite age [27, 28]. In addition, Special Relativity tells us there is a finite maximum speed of information i.e. the speed of light [29]. This means that only points in space separated by less than some finite distance can have had any causal interactions with one another. This critical distance is known as the *particle horizon*.

We can calculate the size of the particle horizon by noting that, in General Relativity, light follows a null geodesic with $ds = 0$. In comoving coordinates, the particle horizon is, therefore, given by

$$\sigma(t) = \int_0^{\sigma(t)} d\sigma = \int_0^t \frac{1}{a} dt = \int_0^{a(t)} \frac{1}{a^2 H} da, \tag{6.28}$$

where

$$d\sigma^2 = \frac{1}{1 - kr^2} dr^2 + r^2 d\theta^2 + r^2 \sin^2\theta d\phi^2 \tag{6.29}$$

is the infinitesimal comoving distance and we have used (6.1).

We can translate this into a physical distance by multiplying by the scale factor. The particle horizon at time t_1 will have a physical size at time t_2 of

$$D_H(t_1, t_2) = a(t_2)\sigma(t_1). \tag{6.30}$$

Note that, by definition, $D_H(t, t)$ is the size of the observable Universe at time t.

Let us compare the physical size of the particle horizon at some early time t with the size of the observable Universe today. We find

$$\frac{D_H(t_0, t_0)}{D_H(t, t_0)} = \frac{\int_0^1 \frac{1}{a^2 H} da}{\int_0^a \frac{1}{a^2 H} da} \approx \frac{(aH)_t}{(aH)_0}. \tag{6.31}$$

During both the radiation and matter dominated epochs, aH is a decreasing function of time and so this ratio is greater than one. In fact we have

$$\frac{D_H(t_0, t_0)}{D_H(t, t_0)} \approx \begin{cases} \frac{1}{a(t)} & t \in \text{radiation domination}, \\ \frac{1}{\sqrt{a(t)}} & t \in \text{matter domination}. \end{cases} \tag{6.32}$$

This means that the observable Universe contains many different regions that were causally disconnected at earlier times in its history.

As an example, let us consider the time of last scattering when the CMB was formed. From observations we know that the surface of last scattering corresponds to $a_{ls} \sim 1/1000$, which is during matter domination, and thus

$$D_H(t_0, t_0) \sim 35 D_H(t_{ls}, t_0). \tag{6.33}$$

Therefore, the CMB that we observe today is made up of approximately 1000 different causally disconnected regions.

Since no information can have been passed between these regions before the CMB was emitted, there is no reason to expect any correlation between the radiation coming from two different regions. From statistical considerations, we should, therefore, expect to find order one differences between them. However, current observations of the CMB show that its temperature is homogeneous to the level of 4.6×10^{-5} [30, 31]. There is no explanation in the HBB model for why these causally disconnected regions of the sky should have such similar temperatures. This is known as the horizon problem.

A more acute version of the horizon problem exists when we consider large scale structure today. The distribution of galaxies today is extraordinarily homogeneous [32]. But, these galaxies grew out of primordial density perturbations that were formed in the very early Universe, long before the time of last scattering [33, 34]. As we have seen, the particle horizon at early times was much smaller than it is today and, therefore, the density perturbations that seeded today's galaxies came from many different causally disconnected regions. These regions should have no correlations and so we should expect an order one difference in the primordial density perturbations and, hence, an order one difference in the number of galaxies in each of these regions.

The exact number of regions depends on when exactly these perturbations were formed. Taking the extreme view that the perturbations were caused by quantum gravity at the Planck time we find

$$\frac{D_H(t_0, t_0)}{D_H(t_P, t_0)} \approx \frac{T_P}{T_0} \sim 10^{31}, \tag{6.34}$$

where we have used the relation (6.22). If the density perturbations were indeed formed at the Planck time then the observable universe is made up of 10^{93} causally disconnected regions. There is no explanation within the HBB theory for why every single one of these regions happened to produce the same magnitude of perturbation.

6.2.3 *The Monopole and Primordial Perturbation Problems*

Many UV completions of the Standard Model involve unifying the standard electromagnetic, weak and strong forces into a single force at high energy [35, 36]. Such Grand Unified Theories (GUTs) predict a symmetry breaking event in the very early Universe when this unity was broken. During this event topological defects can form, the most common of which is a magnetic monopole [37, 38]. In most GUTs these monopoles are stable and extremely massive with $M_{mono} \sim 10^{16}$ GeV. The expectation from the physics of the GUT is that roughly one of these defects should form in each causal region of the Universe.

As we have seen previously, a single causal region from the early Universe takes up a tiny fraction of the observable Universe today. There should, therefore, be of order

$$N_{mono} = \left(\frac{D_H(t_0, t_0)}{D_H(t_P, t_0)}\right)^3 \approx \left(\frac{T_{GUT}}{T_0}\right)^3 \sim 10^{87} \tag{6.35}$$

monopoles in the observable Universe today, assuming a GUT temperature of around 10^{16} GeV. If these monopoles really existed, their mass would be the dominant energy component in the Universe and their density would be far above the critical density needed to cause a Big Crunch: $\rho_{crit} = 3H_0^2$. Since the Universe has not recollapsed we must conclude that the density of monopoles is far less than predicted by the HBB theory.

In fact, after several years of dedicated monopole searches with no observations, the current limits on the flux of magnetic monopoles are

$$\Phi \lesssim 10^{-17} - 10^{-15}\,\text{cm}^{-2}\,\text{s}^{-1}\,\text{sr}^{-1} \tag{6.36}$$

depending on the mass and nature of the monopole [39–41]. This corresponds to a maximum number of monopoles in the observable Universe on the order of

$$N_{mono} \lesssim 10^{63} - 10^{65}, \tag{6.37}$$

far below the expected number from the HBB model.

The final shortcoming of the HBB model that we will discuss here relates again to the primordial density perturbations that seed the creation of large scale structure in the Universe today. The HBB model provides no explanation for the origin of these perturbations, let alone the approximately scale invariant spectrum we infer from measurements of the CMB [31]. In fact, in the HBB model the Universe is exactly homogeneous and so any perturbations must be put in by hand. Therefore, regardless of the status of the other cosmological puzzles, we need to add a mechanism for generating a nearly scale invariant spectrum of primordial density perturbations to the HBB model.

6.3 Inflation

From the above calculations we can see that both the flatness and horizon problems, as well as the monopole problem, arise because the function aH decreases over time. Not only does this decrease cause the density to evolve away from $\Omega = 1$, but it is also the reason why the current Universe contains many causally disconnected regions from earlier times. These problems can, therefore, be solved if there was a period of time in the early Universe during which aH was *increasing*. This is the idea behind the theory of inflation [11–13].

How long does such a period need to last in order to solve the classic cosmological puzzles? We see from (6.27) that if we had

$$\frac{(aH)_0}{(aH)_P} \gtrsim 1 \tag{6.38}$$

then the curvature contribution to the energy budget of the Universe today would be less than its contribution at the Planck time and no fine tuning would be required. Therefore, to solve the flatness problem requires enough inflation to satisfy (6.38).

Similarly, looking at (6.31), we see that condition (6.38) would imply that the entire observable Universe corresponded to a single causal patch at the Planck scale. This would solve the horizon problem since the whole sky would then correspond to a causally connected region and so there would be no mystery as to why the CMB and the distribution of structures in the Universe is so uniform.

Finally, provided the GUT phase transition occurred before the onset of inflation (a realistic scenario in most models), condition (6.38) would imply that (at most) one monopole is formed in the entire observable Universe. This is well within current limits and so the monopole problem is also solved.

We can translate (6.38) into an order-of-magnitude bound on the duration of inflation. Let us assume that inflation starts at some initial time $t_i \sim t_P$ and continues until some final time t_f at which point the Universe immediately reheats to some temperature T_f, which then decays according to (6.22). We shall also take the Hubble constant at the Planck time to be $H_P \sim M_P$. Thus, (6.38) implies

$$N^{(tot)} \equiv \ln\left(\frac{a_f}{a_i}\right) \gtrsim \ln\left(\frac{T_0}{T_f}\frac{M_P}{H_0}\right) \sim 60 - 100, \tag{6.39}$$

where the exact value on the RHS depends on what value is taken for the reheating temperature T_f. The quantity $N^{(tot)}$ is known as the total number of e-foldings of inflation.

We have made several assumptions in obtaining this estimate, most notably the assumption of instant reheating of the Universe at the end of inflation. A detailed calculation of the physics of reheating is beyond the scope of this thesis. We will, therefore, simply quote that more detailed calculations show the total number of e-foldings of inflation required to solve the cosmological puzzles is [42, 43]

$$N^{(tot)} \gtrsim 60. \tag{6.40}$$

6.3.1 The Inflaton

In order to undergo inflation we see that the Universe must be dominated by a form of matter with $w < -1/3$. We have seen one example of such matter in the form of the cosmological constant Λ. However, inflation cannot be driven by a pure cosmological constant since it needs to end at some point.

Instead inflation is believed to be driven by a single[1] scalar field, known as the inflaton. The Universe is, therefore, described by the action

$$S = \int d^4x\sqrt{-g}\left[\frac{1}{2}R + \frac{1}{2}\partial_\mu\phi\partial^\mu\phi - V(\phi)\right], \tag{6.41}$$

where ϕ is the inflaton field and $V(\phi)$ is its potential.

The stress energy tensor for the inflaton field can be calculated from (6.4). If we assume the inflaton field is homogeneous and the Universe is described by the FRW metric (6.1), then we have

$$T_{\mu\nu} = \begin{pmatrix} \frac{1}{2}\dot{\phi}^2 + V & 0 & 0 & 0 \\ 0 & \left(\frac{1}{2}\dot{\phi} - V\right)\frac{a^2}{1-kr^2} & 0 & 0 \\ 0 & 0 & \left(\frac{1}{2}\dot{\phi} - V\right)a^2r^2 & 0 \\ 0 & 0 & 0 & \left(\frac{1}{2}\dot{\phi} - V\right)a^2r^2\sin^2\theta \end{pmatrix}. \tag{6.42}$$

[1] Models with multiple inflaton fields also exist and have had a great amount of development in recent times [44–50]. For the purposes of this chapter and the next, however, we shall focus only on single field inflation. We shall briefly discuss multifield models in Chap. 8.

We see that the inflaton field can be described as a perfect fluid with density

$$\rho = \frac{1}{2}\dot{\phi}^2 + V(\phi) \tag{6.43}$$

and pressure

$$p = \frac{1}{2}\dot{\phi}^2 - V(\phi), \tag{6.44}$$

and, hence, equation of state

$$w = \frac{\dot{\phi} - 2V(\phi)}{\dot{\phi} + 2V(\phi)}. \tag{6.45}$$

It will, therefore, drive inflation whenever $\dot{\phi} < 4V(\phi)$.

During inflation the energy budget of the Universe is dominated by the inflaton field. We can, therefore, plug (6.43) and (6.44) into the Friedmann equations to obtain the evolution equations of the Universe for this period. We find

$$\ddot{\phi} = -3H\dot{\phi} - V'(\phi), \tag{6.46}$$

$$3H^2 = \frac{1}{2}\dot{\phi}^2 + V(\phi), \tag{6.47}$$

where a prime $'$ indicates derivative with respect to ϕ.

6.3.2 Slow Roll Inflation

The equation of motion (6.46) is closely analogous to Newton's second law of motion. Indeed, if the Universe were not expanding (i.e. $H = 0$), then (6.46) would be identical to (5.3) with $V'(\phi)$ acting as the force. The term $-3H\dot{\phi}$ can be interpreted as an additional force proportional to velocity that always acts opposite to the motion and so is analogous to friction. This term is, therefore, known as Hubble friction or Hubble drag.

This Hubble friction term causes the evolution of the inflaton field to slow down, just as physical friction does in classical mechanics. Therefore, the inflaton will eventually enter a regime where

$$\dot{\phi}^2 \ll V(\phi). \tag{6.48}$$

This regime is known as *slow roll* [51–54].

The slow roll regime allows us to make some useful approximations. For example, plugging (6.48) into (6.47) yields

$$H^2 \approx \frac{1}{3}V(\phi). \tag{6.49}$$

Similarly, differentiating (6.48) with respect to time tells us $\ddot{\phi} \ll V'(\phi)$, which simplifies (6.46) to

$$3H\dot{\phi} \approx V'(\phi). \tag{6.50}$$

We can combine (6.49) and (6.50) to derive

$$\dot{\phi} = \sqrt{3}\frac{V'(\phi)}{\sqrt{V(\phi)}}. \tag{6.51}$$

During slow roll the inflaton's equation of state is $w \approx -1$ and so the Hubble parameter H is almost constant. In fact, we see that the relative change of H in a Hubble time is

$$\epsilon_H \equiv \frac{1}{H}\frac{\dot{H}}{H} \approx \frac{3}{2}\frac{\dot{\phi}^2}{V} \ll 1. \tag{6.52}$$

Here, we have defined ϵ_H, which is known as the first slow roll parameter.

Equations (6.49) and (6.50) allow us to derive another expression for ϵ_H, valid only during slow roll. We have

$$\epsilon_H \approx \frac{1}{2}\left(\frac{V'}{V}\right)^2 \equiv \epsilon_V. \tag{6.53}$$

This expression now depends only on the inflaton potential and so is useful for calculations. For example, since inflation ends when $\epsilon_H = 1$, we see that the inflaton field at the end of inflation, ϕ_f, satisfies

$$V'(\phi_f) = \sqrt{2}\, V(\phi_f). \tag{6.54}$$

It is also common to define a second slow roll parameter

$$\eta_H \equiv \frac{1}{H}\frac{\dot{\epsilon}_H}{\epsilon_H}, \tag{6.55}$$

which keeps track of the relative change in ϵ_H during one Hubble time. During slow roll, we can again write this parameter in terms of the inflaton potential and we find

$$\eta_H \approx \frac{V''}{V} \equiv \eta_V. \tag{6.56}$$

In the slow roll regime we have both $\epsilon_H \ll 1$ and $|\eta_H| \ll 1$.

A near constant H implies a near exponential expansion of the Universe during inflation. Indeed, in the limit $\dot{\phi}/V \to 0$ the inflaton behaves indentically to a cosmological constant. It is, therefore, convenient to work with the number of e-foldings

$$N(t) \equiv \ln\left(\frac{a_f}{a(t)}\right). \tag{6.57}$$

Note that with this definition N increases in the opposite direction to time, with earlier times being given by larger values of N.

We can use the slow roll approximations to calculate a useful expression for $N(t)$ with the following manipulation

$$N(t) = \int_{a(t)}^{a_f} \frac{1}{a} da = \int_{\phi(t)}^{\phi_f} \frac{H}{\dot{\phi}} d\phi \approx \int_{\phi(t)}^{\phi_f} \frac{V(\phi)}{V'(\phi)} d\phi. \tag{6.58}$$

By comparing (6.58) with (6.40) we see that a period of slow roll inflation is sufficient to solve the flatness, horizon and monopole problems provided

$$\int_{\phi_i}^{\phi_f} \frac{V(\phi)}{V'(\phi)} d\phi \gtrsim 60, \tag{6.59}$$

where ϕ_i is the initial value of the inflaton field at the onset of slow roll.

For many models, the condition (6.59) forces us to consider inflaton field values that are $\phi \gg M_P$. One may worry that such large field values requires us to use the full UV-complete theory of quantum gravity. However, even though the field strength is large, the value of the inflationary potential typically remains sub-Planckian. Corrections from quantum gravity will, therefore, be small during the inflationary era.

Nonetheless, there is some evidence that the large displacement in field space required by (6.59) is incompatible with string theory and other leading theories of quantum gravity [55]. This has been used by some to suggest that the theory of inflation is in tension with UV physics [56–58]. These claims are still at the level of conjecture, however. Therefore, while we need to be aware of such issues, we should not rule out inflation on the basis of these claims until the underlying conjectures are proven.

We have so far assumed the inflaton field to be completely homogeneous and isotropic in accordance with the postulates of leading order cosmology. However, this homogeneity will not be exact and the inflaton will pick up perturbations. An in-depth calculation of the spectrum of these perturbations is beyond the scope of this thesis and so we will simply state without proof that this spectrum is very nearly scale invariant [59] and matches the current observations of the CMB [60]. Thus, inflation solves the primordial perturbation problem as well.

6.4 The Initial Conditions of Inflation

We have seen how an extended period of slow-roll inflation can solve the classic cosmological puzzles of the HBB model. However, it is important to point out that these puzzles are fundamentally issues of fine tuning. There is nothing in the HBB model to prevent the Universe from starting out in a completely flat, completely homogeneous state with no monopoles and a scale-invariant spectrum of density perturbations. It is just that this scenario would require an extreme fine tuning of the initial conditions of the Universe.

This means, however, that if inflation is to be a real solution to these problems, it cannot have any fine tuning issues of its own. If inflation itself required finely tuned initial conditions then we would not be solving the fine tuning problems of the HBB model, but merely pushing them back to an earlier time. In other words, for inflation to be a successful solution to the cosmological puzzles, it must require less fine tuning than simply setting the initial conditions of the HBB model to acceptable values "by hand".

In order to be concrete, let us clarify what we mean by the initial conditions of inflation. The Eqs. (6.46) and (6.47) were derived from classical General Relativity. However, we know that far enough back in the history of the Universe, the temperature was so high ($T \gtrsim M_P$) that quantum corrections to these equations were no longer negligible. Although there has been much progress in developing a UV-complete quantum theory of gravity [61–68], no consensus has yet been reached and it is still unclear how the Universe evolved before this point. The quantum gravity regime is, therefore, often treated as a "black box" that spits the Universe out in some state, whose subsequent evolution can be well described by QFT and General Relativity. It is this state that we refer to as the initial conditions of inflation.

The question of initial conditions of inflation can be broken down into two parts. First, there is no guarantee that the Universe will exit the quantum gravity regime in a homogeneous state and so we need to understand how inflation responds to initial inhomogeneities. Second, even if those inhomogeneities are negligible, then we need to know the initial values of the inflaton field and its derivative in order to solve the Friedmann equations. We should, therefore, endeavour to understand how sensitive the predictions of this chapter are to both the initial inhomogeneities and the initial field values in order to determine whether inflation requires finely tuned initial conditions.

6.4.1 Initial Inhomogeneities

It has long been argued that inflation is very robust to initial inhomogeneities [69–77]. The argument, most recently described by Linde [78], stems from the fact that "normal" forms of matter (i.e. non-relativistic matter, curvature and radiation) dilute as the Universe expands, whereas the potential energy of the inflaton field does not. Therefore, even if there are some local regions that gravitationally collapse, provided the Universe continues to expand overall, this potential energy will eventually dominate leading to the onset of inflation.

However, there are some [79–82] that disagree with these conclusions, arguing that the dynamics of the inflaton field are far more complicated. Unlike a true cosmological constant, the inflaton potential can change its value. The inflaton field has non-trivial interactions with the gravitational background and it is argued that these back-reaction effects can be important.

Recent developments in numerical relativity [83–85] have shown that which effect wins out is strongly dependent on the shape of the inflationary potential. Large field models of inflation (in which the inflaton field traverses a distance in field space $\delta\phi > M_p$) and those for which the potential is convex ($V''(\phi) > 0$ in the inflationary part of the potential), have been shown to be extremely robust. In these models any initial inhomogeneities are washed out long before the onset of slow roll inflation.

Models with both small field excursions and concave potentials, on the other hand, have been shown to be less robust. For these models, inhomogeneities can cause inflation to end prematurely and, hence, significantly less inflation can occur than is predicted by (6.58). This can happen even if the energy density of the spatial gradient of the inflaton field is subdominant to its potential energy [85].

It is often argued, however, that these observations are irrelevant because of anthropic considerations. In a large Universe, different regions of space will have different initial conditions. Some of these regions will have initial conditions that allow inflation and some will not. Regions that allow for inflation grow exponentially whereas the regions that do not allow inflation will grow only polynomially. The argument, therefore, says that no matter how uncommon the regions that allow for inflation are, they will eventually come to dominate the volume of the Universe and so we are most likely to find ourselves in one of these regions. This argument is known as volume weighting because it assumes that the likelihood of existing in a particular region of space is proportional to its current volume [86–92].

Although volume weighting seems like a sensible thing to do, it often leads to problems. For example, in most theories of cosmology, the Universe is infinite in extent. Since it is impossible to have a uniform distribution on an infinite space, the assumption behind volume weighting—that we are equally likely to be at each point in space—breaks down. Furthermore, generically in these cases both the inflationary and non-inflationary patches take up infinite volume [93, 94]. The ratio of these volumes is, therefore, ambiguous and it is impossible to say which region is "bigger".

Volume weighting also leads to the so-called *youngness paradox* [95, 96]. Inflation does not stop in all regions at the same time and, in most models, there are some regions of the Universe that keep inflating forever (this is known as eternal inflation [97, 98]). In such models new universes are being born all the time as inflation ends in another region of space. Because of the exponential nature of inflation these new universes outnumber our own by an enormous factor. For example, patches that stopped inflating one second later than our own outnumber us by $e^{10^{37}}:1$. Therefore, if we employ volume weighting, we would be exponentially more likely to find ourselves in one of these younger patches and it is very surprising that our Universe is so old. Even if we employ anthropic reasoning, it is difficult to argue that intelligent life could not have evolved even one second earlier than it did.

6.4.2 Initial Field Values

Let us now focus on the second aspect of the initial conditions of inflation: the initial values of the inflaton field and its derivative. These values will serve as the initial conditions for (6.46) and (6.47) and are, therefore, essential in understanding the evolution of the early Universe.

The initial field values are often overlooked in the literature. This is because in a flat FRW Universe with a single inflaton field, the slow roll regime is a universal attractor in phase space [51–54]. We can see how this works by considering Fig. 6.1. Here, we have numerically solved the equations of motion (6.46) and (6.47) for a range of different initial conditions, assuming an inflationary potential $V(\phi) = \phi^4$. We see that regardless of the initial position in phase space, the inflaton very quickly evolves to follow the slow roll solution given by (6.51) (shown in red).[2]

The attractor nature of the slow roll regime is both a blessing and a curse. On the one hand, the slow roll attractor allows us to make robust predictions from the inflationary potential alone, independently[3] of the unknown initial conditions. This has been extremely useful in differentiating between models of inflation and has allowed us to hone in on the inflationary potential that best describes our Universe [99]. On the other hand, the slow roll attractor washes out any observable signature of the state of the Universe before inflation. This makes it impossible to infer anything about the quantum gravity era of the Universe by looking at signatures in the CMB [100].

However, the slow roll attractor is not the end of the story of initial conditions. The plot in Fig. 6.1 has been produced assuming a flat universe with $k = 0$. For a universe with non-zero curvature $k \neq 0$, the slow roll attractor is not universal and there are some regions of phase space which never reach it [53, 54]. Furthermore, even if $k = 0$, we can see that there are some trajectories that reach the attractor with $\phi < \phi_f$ and so do not undergo inflation after all.

In fact, in order to solve the flatness, horizon and monopole problems, it is not enough to start slow roll with $\phi > \phi_f$. We have seen that solving these problems requires at least $N \gtrsim 60$ e-foldings of inflation. This means we actually require slow roll to begin with $\phi > \phi_c$, where ϕ_c is some critical value that gives $N = 60$. This critical value depends on the inflationary potential and can be calculated from (6.59). For example, for the potential $V(\phi) = \phi^4$, we have $\phi_c = 22.09M_P$. So we see that, in fact, none of the trajectories shown in Fig. 6.1 would provide enough inflation to solve the cosmological puzzles.

There is one more reason to consider the initial field values of the inflaton. While the example potential $V(\phi) = \phi^4$ is very simple, many realistic models of inflation have far more complicated potentials with different distinct regions. For example, the quintessential model of inflation [101–104] has a potential with two plateaus of different heights separated by a region where the potential changes rapidly. Similarly, the α-attractor model of inflation [105, 106] has a potential with two plateaus of the

---

[2] We can also see that the inflaton eventually deviates from the slow roll trajectory once $\phi < \phi_f$ and inflation ends.

[3] Or, rather, almost independently as we shall see below.

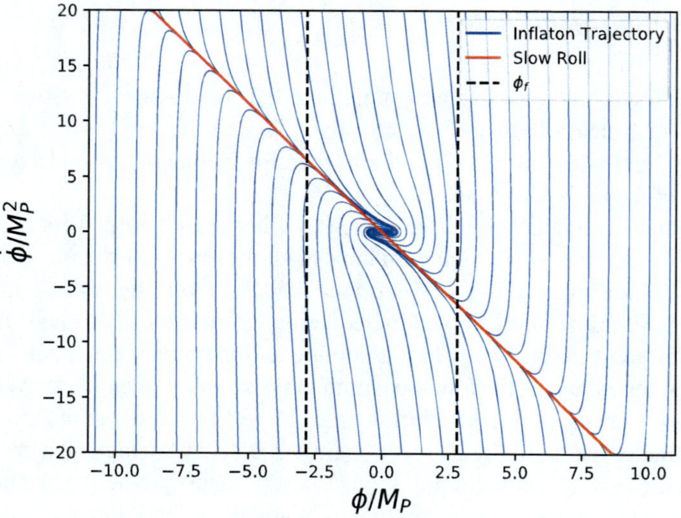

Fig. 6.1 The trajectory of the inflaton field in phase space for a range of different initial conditions, showing the attractor nature of the slow roll regime. Also shown (in red) is the slow roll trajectory given by (6.49) and (6.50). In addition, we have placed a vertical dashed line at $\phi_f \approx 2.82 M_P$, which is the value at which inflation ends. To produce this plot we have assumed a flat Universe with $k = 0$ and an inflationary potential $V(\phi) = \phi^4$

Fig. 6.2 Schematic illustration of the potentials for the α-attractor (dashed blue) and quintessential (solid orange) models of inflation

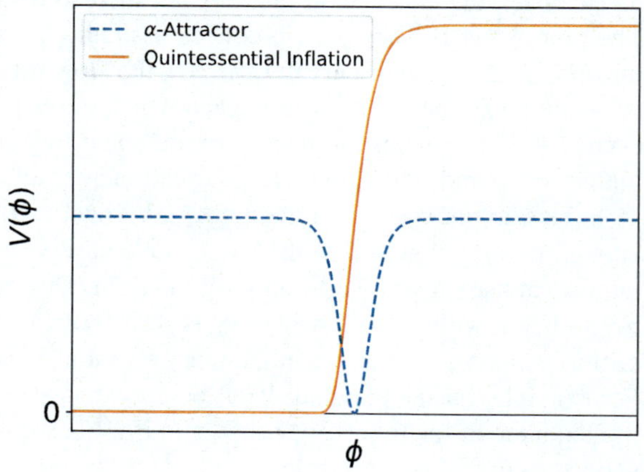

same height separated by a dip. These potentials are shown schematically in Fig. 6.2. For such models, the predictions made depend on which region the inflaton is in during inflation. Inflation from the upper plateau of the quintessential inflation model is consistent with observations, whilst inflation from the lower plateau is not. The region that is relevant for inflationary observables is set by the initial field value.

Determining what initial field values we expect for the inflaton is a difficult problem, bordering on philosophy [107]. In the absence of a UV-complete quantum theory of gravity, the expected distribution of initial field values cannot be predicted from any known theory of physics and must, therefore, be chosen based on philosophical principles such as naturalness and minimalism. A common principle to appeal to in this case is Laplace's principle of indifference [108]. This principle states that, in the absence of any knowledge of a system, every possibility should be considered equally likely.

However, the situation is not that simple. The value of the inflaton field and its derivative are continuous and so it is impossible to evaluate the number of possibilities in a particular region of phase space without some kind of counting measure. What looks like a uniform distribution according one measure will, in general, look like a highly non-trivial distribution according to another. The question of which measure is the most appropriate to use for inflation is still unresolved and is known as the *measure problem* in inflationary cosmology.

As an example, we may be tempted to consider a measure $d\Omega = d\phi d\dot{\phi}$ and correspondingly choose a uniform distribution on the $\phi - \dot{\phi}$ plane. However, the arguments of Chap. 1 tell us that there is nothing special about ϕ and we could just as easily describe inflation in terms of some other field $\varphi = \varphi(\phi)$. If we worked with φ and $\dot{\varphi}$ we would be more inclined to define the measure $d\tilde{\Omega} = d\varphi d\dot{\varphi}$ and, therefore, consider a uniform distribution on the $\varphi - \dot{\varphi}$ plane instead. In general $d\Omega \neq d\tilde{\Omega}$ and so we have broken the cardinal rule of reparametrisation invariance. We have chosen a distribution that depends on the way we choose to write it down.

There is a further issue with this distribution; it is not well defined. Both ϕ and $\dot{\phi}$ are, in general, unbounded and, thus, the total measure is infinite. A uniform distribution on an infinite space makes no sense since it cannot be normalized. In using this measure we have, therefore, implicitly assumed some form of regularisation, for example a cutoff in $|\phi|$ and $|\dot{\phi}|$. Performing this regularisation in a reparametrisation invariant way is highly non-trivial as we shall see.

Despite these issues, such parametrisation dependent distributions continue to be used either implicitly (e.g. when appealing to the slow roll attractor) or explicitly [78, 109, 110] to justify inflation as the solution to the cosmological puzzles. However, as we have seen, their use in the study of initial conditions is highly suspect since it relies heavily on an arbitrary choice of parametrisation and regularisation scheme. We should, therefore, endeavour to use a more natural measure to study the initial conditions of inflation.

One of the first attempts to construct such a natural measure was made by Gibbons, Hawking and Stewart (GHS) [111]. They came up with three conditions, which they felt such a measure should satisfy. These were:

1. The measure should be positive definite.
2. The measure should be reparametrisation invariant.
3. The measure should respect the symmetries of the theory and should not make use of any ad-hoc structures not present in the theory itself.

The measure that they constructed was based on the Liouville measure for Hamiltonian systems [112], which we review briefly here. In a system described by Hamiltonian mechanics (such as homogeneous inflation) each coordinate q^A comes with a conjugate momentum

$$p_A = \frac{\partial L}{\partial \dot{q}^A}, \tag{6.60}$$

where L is the Lagrangian. We can use these conjugate variables to construct a symplectic form on the phase space

$$\omega = \sum_{A=1}^{n} dp_A \wedge dq^A, \tag{6.61}$$

where n is the number of degrees of freedom in the system and $\wedge$ denotes the wedge product. We can use this two-form to construct a measure satisfying properties 1, 2 and 3 above:

$$d\Omega_L = \frac{(-1)^{n(n-1)/2}}{n!}\omega^n = d^n p d^n q. \tag{6.62}$$

This measure is known as the Liouville measure.

However, there are two issues that prevented GHS from simply applying (6.62) to inflation. The first is the Hamiltonian constraint (6.47). We see that this equation implies a relation between the initial values of a, $\dot{a}$, ϕ and $\dot{\phi}$. Therefore, we are not free to choose initial conditions anywhere on the phase space, but only on a restricted region of the phase space that satisfies

$$\mathcal{H} = -3a\dot{a}^2 + \frac{1}{2}a^3\dot{\phi}^2 + a^3 V(\phi) = 0, \tag{6.63}$$

where $\mathcal{H} = p_A q^A - L$ is the Hamiltonian of the theory (6.41). We call this region the Hamiltonian hypersurface.

The second issue stems from the fact that multiple points in phase space correspond to the same solution of the equations of motion, just offset by some constant time. Since the absolute value of time is unobservable, GHS argued that all such points should be identified and, thus, we should define our distribution over the different *trajectories* in phase space rather than just the initial conditions. This requires us to choose a *counting surface* through which each trajectory passes once and only once.

It turns out that both of these issues can be solved by noticing that t and $\mathcal{H}$ are canonical variables for a Hamiltonian system. Therefore, we can rewrite (6.61) as

$$\omega = \sum_{A=1}^{n-1} dp_A \wedge dq^A + d\mathcal{H} \wedge dt. \tag{6.64}$$

We can define a measure over trajectories by restricting ourselves to a counting hypersurface $\mathcal{C}$ on which $\mathcal{H} = 0$ and $t = const$. We see there is a natural two form on this hypersurface $\omega_{\mathcal{C}} = \omega|_{d\mathcal{H}=dt=0}$, which leads to the measure

$$d\Omega_{GHS} = \frac{(-1)^{(n-1)(n-2)/2}}{(n-1)!}\omega_{\mathcal{C}}^{n-1}. \tag{6.65}$$

According to Darbaux's theorem [113], the two-form (6.61) is preserved by the Hamiltonian flow. The Liouville measure (6.62) for a set of trajectories is, therefore, preserved over time in accordance with Liouville's theorem [112, 114]. Thus, the GHS measure (6.65) should give the same results regardless of the choice of counting hypersurface.

In their original paper [111], GHS applied this measure to the theory of inflation described by (6.41). They chose a counting hypersurface corresponding to some early time and calculated what fraction of the trajectories went on to undergo inflation according to the measure (6.65). Although they found the total measure to be infinite, the infinity corresponds to the unbounded range of the scale factor a, whose absolute value is unobservable. This led them to employ a natural regularisation scheme in which a simple cutoff was placed on the value of a. They found that, regardless of the value of this cutoff, the overwhelming majority of trajectories did indeed go on to inflate and, therefore, concluded that inflation did not require finely tuned initial conditions.

However, a subsequent analysis by Gibbons and Turok (GT) [115][4] found conflicting results using the same measure. They took a more Bayesian approach, choosing a counting surface corresponding to today and calculated the fraction of trajectories that underwent inflation in the past. Again they found that the total measure was infinite and so employed the same regularisation scheme as GHS. However, they found that trajectories that underwent N e-folds of inflation were suppressed by a factor e^{-3N}, in complete contradiction to the work of GHS.

The reason for this discrepancy was first noted by Hawking, Page and Chmielowski [117–119] and was later formalised by Remmen and Carroll [120] and Corichi and Sloan [121]. They found that even though the unregulated measure (6.65) is independent of the counting hypersurface, by Liouville's theorem, the same is not true when the measure is regularised. In fact, imposing a cutoff in a on two different counting surfaces corresponds to two, incompatible regularisation schemes.

It is relatively easy to understand why this arises. As we have already seen, trajectories in the $\phi - \dot{\phi}$ plane converge at late times due to the slow roll attractor. But, Liouville's theorem states that the total measure taken up by these trajectories does not change over time. Therefore, if these trajectories are converging in one direction, they must be diverging in another, namely the direction corresponding to a. This means that imposing a cutoff in a at late times removes many more inflationary trajectories than imposing the same cutoff at early times. This explains the difference between the results of GHS and GT.

However, even without the ambiguities of infinite measures, the previous approach has another conceptual problem. GT approached the likelihood of inflation from a Bayesian standpoint, calculating the probability that inflation occurred in the past, given that the Universe looks the way it does today. Whilst Bayesian methods are often the correct way to evaluate probabilities, their use in cosmology is a bit suspect

[4] See also [116] for an extension of this work to multifield models of inflation.

as the following argument by Schiffrin and Wald [122] shows (see also [123] for further discussion).

When considering the Universe as a whole, we know that the past and the future look very different. Due to the second law of thermodynamics, the past Universe was in a low entropy state, whereas the future will be in a high entropy state. However, the equations describing the evolution of the Universe are time-reversal symmetric. Therefore, any attempt to predict the state of the Universe at some time other than today using Bayesian methods will inevitably give the same results regardless of whether that time is in the past or the future. In particular, if we try to *retrodict* the past state of the Universe using these methods we incorrectly conclude that there was a large amount of entropy in the early Universe. The correct state, which necessarily had a low entropy, will be assigned a very low Bayesian probability.

In light of these observations we propose extending the list above with two more requirements for a measure on the initial conditions:

4. The measure should either be finite or be independent of the method of regularisation.
5. The probability of inflation should be calculated using frequentist, not Bayesian, techniques.

In the following chapter we shall see how we can use the techniques developed throughout this thesis to construct a measure that obeys both of these new requirements in addition to the requirements of GHS.

References

1. Einstein A (1915) The field equations of gravitation. In: Sitzungsber Preuss Akad Wiss Berlin (Math Phys), pp 844–847 (cit on pp 19, 39, 138, 174)
2. Groot Nibbelink S, van Tent BJW (2000) Density perturbations arising from multiple field slow roll inflation. arXiv:hep-ph/0011325 [hep-ph] (cit on pp 51, 173)
3. Gong J-O, Tanaka T (2012) A covariant approach to general field space metric in multi-field inflation. JCAP 1103 (2011). [Erratum: JCAP1202,E01(2012)], p 015. arXiv:1101.4809 [astro-ph.CO] (cit on pp 51, 173)
4. Friedmann A (1922) Über die Krümmung des Raumes. In: Zeitschrift für Physik 10.1, pp 377–386 (cit on pp 82, 173)
5. Friedmann A (1924) Über die Möglichkeit einer Welt mit konstanter negative Krümmung des Raumes. In: Zeitschrift für Physik 21.1 (1924), pp 326–332 (cit on pp 82, 173)
6. Lemaître AG (1931) A homogeneous universe of constant mass and increasing radius accounting for the radial velocity of extra-galactic Nebulæ. In: Monthly notices of the royal astronomical society 91.5, pp 483–490 (cit on pp 82, 173)
7. Robertson HP (1935) Kinematics and world-structure. Astrophys J 82, 284 (cit on pp 82, 173)
8. Robertson HP (1936) Kinematics and world-structure II. Astrophys J 83, p 187 (cit on pp 82, 173)
9. Robertson HP (1936) Kinematics and world-structure III. Astrophys J 83, 257 (cit on pp 82, 173)
10. Walker AG (1937) On Milne's theory of world-structure*. In: Proceedings of the London mathematical society s2-42.1, pp 90–127 (cit on pp 82, 173)

11. Guth AH (1981) The inflationary universe: a possible solution to the horizon and flatness problems. Phys Rev D23:347–356 (cit on pp 154, 184, 249)
12. Linde AD (1982) A new inflationary universe scenario: a possible solution of the horizon, flatness, homogeneity, isotropy and primordial monopole problems. Phys Lett 108B, 389–393. (cit on pp 154, 184, 249)
13. Albrecht A, Steinhardt PJ (1982) Cosmology for grand unified theories with radiatively induced symmetry breaking. Phys Rev Lett 48:1220–1223 (cit on pp 154, 184, 249)
14. Groot Nibbelink S, van Tent BJW (2002) Scalar perturbatslow-roll inflation. In: Class Quant Grav 19:613–640. arXiv:hep-ph/0107272 [hep-ph] (cit on p 173)
15. Gong J-O, Stewart ED (2002) The power spectrum for a multicomponent inflaton to second order corrections in the slow roll expansion. Phys Lett B538:213–222. arXiv:astro-ph/0202098 [astro-ph] (cit on p 173)
16. van Tent BJW (2004) Multiple-field inflation and the cmb. Class Quant Grav 21:349–370 (2004). arXiv:astro-ph/0307048 [astro-ph] (cit on p 173)
17. Langlois D, Renaux-Petel S (2008) Perturbations in generalized multi-field inflation. JCAP 0804:017. arXiv:0801.1085 [hep-th] (cit on p 173)
18. Gorbunov DS, Rubakov VA (2011) Introduction to the theory of the early universe: hot big bang theory. World Scientific (cit on pp 174, 216)
19. Raychaudhuri A (1955) Relativistic cosmology. 1. Phys Rev 98:1123–1126 (cit on p 176)
20. Penzias AA, Wilson RW (1965) A Measurement of excess antenna temperature at 4080-Mc/s. Astrophys J 142:419–421 (cit on pp 178, 179)
21. Planck M (1914) The theory of heat radiation. Blakiston (cit on p 178)
22. Aghanim N et al (2018) Planck 2018 results. VI. Cosmological parameters. arXiv:1807.06209. [astro-ph.CO] (cit on pp 179, 180, 226, 233)
23. Hubble E (1929) A relation between distance and radial velocity among extra-galactic nebulae. Proc Nat Acad Sci 15, 168–173 (cit on p 179)
24. Yang J-M, Turner MS, Steigman G, Schramm DN, Olive KA (1984) Primordial nucleosynthesis: a critical comparison of theory and observation. Astrophys J 281, 493–511 (cit on p 179)
25. Cyburt RH, Fields BD, Olive KA, Yeh T-H (2016) Big bang nucleosynthesis: 2015. Rev Mod Phys 88:015004. arXiv:1505.01076 [astro-ph.CO] (cit on p 179)
26. Alam S et al (2017) The clustering of galaxies in the completed SDSS-III Baryon Oscillation Spectroscopic Survey: cosmological analysis of the DR12 galaxy sample. Mon Not Roy Astron Soc 470.3:2617–2652. arXiv:1607.03155 [astro-ph.CO] (cit on p 180)
27. Hawking SW (1966) Properties of expanding universes. PhD thesis. Cambridge U (cit on p 181)
28. Penrose R (1965) Gravitational collapse and space-time singularities. Phys Rev Lett 14, 57–59 (cit on p 181)
29. Einstein A (1905) Zur Elektrodynamik bewegter Körper". In: Annalen der Physik vol 322, 10, pp 891–921 (cit on p 181)
30. Akrami Y et al (2018) Planck 2018 results. I. Overview and the cosmological legacy of Planck. arXiv:1807.06205 [astro-ph.CO] (cit on p 182)
31. Aghanim N et al (2019) Planck 2018 results. V. CMB power spectra and likelihoods (2019). arXiv:1907.12875 [astro-ph.CO] (cit on pp 182, 184)
32. Ahumada R et al (2020) The sixteenth data release of the sloan digital sky surveys: first release from the APOGEE-2 southern survey and full release of eBOSS spectra. Astrophys J Suppl 249.1:3. arXiv:1912.02905 [astro-ph.GA] (cit on p 182)
33. Peebles PJE (1980) The large-scale structure of the universe. Princeton series in physics. Princeton University Press (cit on p 182)
34. Padmanabhan T (1993) Structure formation in the universe. Cambridge University Press (cit on p 182)
35. Georgi H, Glashow SL (1974) Unity of all elementary particle forces. Phys Rev Lett 32, 438–441 (cit on p 183)

36. Bertolini S, Di Luzio L, Malinsky M (2009) Intermediate mass scales in the non-supersymmetric SO(10) grand unification: a reappraisal. Phys Rev D 80:015013. arXiv:0903.4049 [hep-ph] (cit on p 183)
37. 't Hooft G (1974) Magnetic monopoles in unified gauge theories. In: Taylor JC (ed) Nuclear Physics B vol 79, pp 276–284 (cit on p 183)
38. Polyakov AM (1974) Particle spectrum in the quantum field theory. In: Taylor JC (ed)JETP letters vol 20, pp 194–195 (cit on p 183)
39. Patrizii L, Spurio M (2015) Status of searches for magnetic monopoles. Ann Rev Nucl Part Sci 65:279–302. arXiv:1510.07125 [hep-ex] (cit on p 184)
40. Giacomelli G, Giorgini M, Lari T et al (2000) Magnetic monopole bibliography, May 2000. arXiv:hep-ex/0005041 (cit on p 184)
41. Balestra S, Giacomelli G, Giorgini M et al (2011) Magnetic monopole bibliography-II, May 2011. arXiv:1105.5587 [hep-ex] (cit on p 184)
42. Liddle AR, Leach SM (2003) How long before the end of inflation were observable perturbations produced? Phys Rev D 68:103503. arXiv: astro-ph/0305263 (cit on p 185)
43. Dodelson S, Hui L (2003) A horizon ratio bound for inflationary fluctuations. Phys Rev Lett 91:131301. arXiv: astro-ph/0305113 (cit on p 185)
44. Linde AD (1994) Hybrid inflation. In: Phys Rev D 49:748–754. arXiv:astro-ph/9307002 (cit on p 186)
45. Conlon JP, Quevedo F (2006) Kahler moduli inflation. JHEP 01:146. arXiv:hep-th/0509012 (cit on p 186)
46. Bond JR, Kofman SP, Vaudrevange PM (2007) Roulette inflation with Kahler moduli and their axions. Phys Rev D 75:123511. arXiv:hep-th/0612197 (cit on p 186)
47. Langlois D, Renaux-Petel S, Steer DA, Tanaka T (2008) Primordial perturbations and non-Gaussianities in DBI and general multi-field inflation. Phys Rev D78, 063523. arXiv:0806.0336 [hep-th] (cit on p 186)
48. Christodoulidis P, Roest D, Sfakianakis EI (2019) Angular inflation in multi-field α-attractors. JCAP 11:002. arXiv:1803.09841 [hep-th] (cit on p 186)
49. Christodoulidis P, Roest D, Sfakianakis EI (2019) Scaling attractors in multi-field inflation. JCAP 12:059. arXiv:1903.06116 [hep-th] (cit on p 186)
50. Dimakis N, Paliathanasis A, Terzis PA, Christodoulakis T (2019) Cosmological solutions in multiscalar field theory. Eur Phys J C 79.7:618. arXiv:1904.09713 [gr-qc] (cit on p 186)
51. Albrecht A, Brandenberger RH, Matzner R (1987) Inflation with generalized initial conditions. Phys Rev D35:429 (cit on pp 187, 193)
52. Brandenberger RH, Feldman H, Kung J (1991) Initial conditions for chaotic inflation. Phys Scripta T36:64–69 (cit on pp 187, 193)
53. Salopek DS, Bond JR (1990) Nonlinear evolution of long wavelength metric fluctuations in inflationary models. Phys Rev D42:3936-3962 (cit on pp 187, 193, 237)
54. Liddle AR, Parsons P, Barrow JD (1994) Formalizing the slow roll approximation in inflation. Phys Rev D50:7222–7232. arXiv:astro-ph/9408015 [astro-ph] (cit on pp 187, 193, 237)
55. Ooguri H, Vafa C (2007) On the geometry of the string landscape and the swampland. Nucl Phys B 766:21–33. arXiv: hep-th/0605264 (cit on pp 189, 252)
56. Garg SK, Krishnan C (2019) Bounds on slow roll and the de sitter Swampland. JHEP 11:075. arXiv:1807.05193 [hep-th] (cit on p 189)
57. Agrawal P, Obied G, Steinhardt PJ, Vafa C (2018) On the cosmological implications of the String Swampland. Phys Lett B 784:271–276. arXiv:1806.09718 [hep-th] (cit on p 189)
58. Brandenberger R (2019) Fundamental physics, the Swampland of effective field theory and early universe cosmology. In: 11th international symposium on quantum theory and symmetries, Nov 2019. arXiv:1911.06058 [hep-th] (cit on p 189)
59. Gorbunov DS, Rubakov VA (2011) Introduction to the theory of the early universe: cosmological perturbations and inflationary theory. v 2. World Scientific (cit on p 190)
60. Akrami Y et al (2018) Planck 2018 results. X constraints on inflation. arXiv:1807.06211 [astro-ph.CO] (cit on pp 190, 233, 252)

61. Rovelli C (2000) Notes for a brief history of quantum gravity. In: 9th Marcel Grossmann meeting on recent developments in theoretical and experimental general relativity, gravitation and relativistic field theories (MG 9), June 2000, pp 742–768. arXiv: gr-qc/0006061 (cit on p 190)
62. Carlip S, Chiou D-W, Ni W-T, Woodard R (2015) Quantum gravity: a brief history of ideas and some prospects. Int J Mod Phys D 24.11:1530028. arXiv:1507.08194 [gr-qc] (cit on p 190)
63. Witten E (1995) String theory dynamics in various dimensions. Nucl Phys B 443, 85–126 (1995). arXiv: hep-th/9503124 (cit on p 190)
64. Polchinski J (1998) String theory. Cambridge Monographs on Mathematical Physics, vol 1. Cambridge University Press (cit on p 190)
65. Gambini R, Pullin J (2020) Loop quantum gravity for everyone. World Scientific (cit on p 190)
66. Ashtekar A (1986) New variables for classical and quantum gravity. Phys Rev Lett 57:2244–2247 (cit on p 190)
67. Ashtekar A (1987) New Hamiltonian formulation of general relativity. Phys Rev D 36:1587–1602 (cit on p 190)
68. Morris TR (2018) Perturbatively renormalizable quantum gravity. Int J Mod Phys D 27.14:1847003. arXiv:1804.03834 [hep-th] (cit on p 190)
69. Barrow JD, Stein-Schabes J (1984) Inhomogeneous cosmologies with cosmological constant. Phys Lett A103:315–317 (cit on p 191)
70. Linde AD (1985) Initial conditions for inflation. Phys Lett B 162:281–286 (cit on p 191)
71. Jensen LG, Stein-Schabes JA (1987) Is inflation natural? Phys Rev D35:1146 (cit on p 191)
72. Goldwirth DS, Piran T (1990) Inhomogeneity and the onset of inflation. Phys Rev Lett 64:2852–2855 (cit on p 191)
73. Goldwirth DS (1991) On inhomogeneous initial conditions for inflation. Phys Rev D43:3204–3213 (cit on p 191)
74. Goldwirth DS, Piran T (1992) Initial conditions for inflation. Phys Rept 214:223–291 (cit on p 191)
75. Bruni M, Mena FC, Tavakol RK (2002) Cosmic no hair: nonlinear asymptotic stability of de Sitter universe. Class Quant Grav 19:L23-L29. arXiv: gr-qc/0107069 [gr-qc] (cit on p 191)
76. East WE, Kleban M, Linde A, Senatore L (2016) Beginning inflation in an inhomogeneous universe. JCAP 1609.09:010. arXiv:1511.05143 [hep-th] (cit on p 191)
77. Kleban M, Senatore L (2016) Inhomogeneous Anisotropic Cosmology. JCAP 1610.10:022. arXiv:1602.03520 [hep-th] (cit on p 191)
78. Linde A (2018) On the problem of initial conditions for inflation. Found Phys 48.10:1246–1260. arXiv:1710.04278 [hep-th] (cit on pp 191, 196)
79. Penrose R (1989) Difficulties with inflationary cosmology. Ann N Y Acad Sci 571:249–264 (cit on p 191)
80. Ijjas A, Steinhardt PJ, Loeb A (2013) Inflationary paradigm in trouble after Planck 2013. Phys Lett B723:261–266. arXiv:1304.2785 [astro-ph.CO] (cit on p 191)
81. Ijjas A, Steinhardt PJ, Loeb A (2014) Inflationary schism. Phys Lett B736:142–146. arXiv:1402.6980 [astro-ph.CO] (cit on p 191)
82. Berezhiani L, Trodden M (2015) How Likely are constituent quanta to initiate inflation? Phys Lett B749:425–430. arXiv:1504.01730 [hep-th] (cit on p 191)
83. Clough K, Lim EA, DiNunno BS et al (2017) Robustness of inflation to inhomogeneous initial conditions. JCAP 1709.09:025. arXiv:1608.04408 [hep-th] (cit on p 191)
84. Clough K, Flauger R, Lim EA (2018) Robustness of inflation to large tensor perturbations. JCAP 1805.05:065. arXiv:1712.07352 [hep-th] (cit on p 191)
85. Aurrekoetxea JC, Clough K, Flauger R, Lim EA (2020) The effects of potential shape on inhomogeneous inflation. JCAP 05:030. arXiv:1910.12547 [astro-ph.CO] (cit on pp 191, 192)
86. Linde AD (1986) Eternally existing selfreproducing chaotic inflationary universe. Phys Lett B 175:395–400 (cit on p 192)

87. Linde AD, Mezhlumian A (1993) Stationary universe. Phys Lett B307:25–33. arXiv:gr-qc/9304015 [gr-qc] (cit on p 192)
88. Garriga J, Schwartz-Perlov D, Vilenkin A, Winitzki S (2006) Probabilities in the inflationary multiverse. JCAP 01:017. arXiv:hep-th/0509184 (cit on p 192)
89. Easther R, Lim EA, Martin MR (2006) Counting pockets with world lines in eternal inflation. JCAP 03:016. arXiv:astro-ph/0511233 (cit on p 192)
90. Vilenkin A (2007) A measure of the multiverse. J Phys A40:6777. arXiv:hep-th/0609193 [hep-th] (cit on p 192)
91. Aguirre A, Gratton S, Johnson MC (2007) Hurdles for recent measures in eternal inflation. Phys Rev D 75:123501. arXiv:hep-th/0611221 (cit on p 192)
92. Clifton T, Shenker S, Sivanandam N (2007) VolumeWeighted measures of eternal inflation in the Bousso-Polchinski landscape. JHEP 09:034. arXiv:0706.3201 [hep-th] (cit on p 192)
93. Vilenkin A (1983) Birth of inflationary universes. Phys Rev D27:2848–2855 (cit on p 192)
94. Brandenberger R (2016) Initial conditions for inflation—a short review. Int J Mod Phys D26.01:1740002. arXiv:1601.01918 [hep-th] (cit on p 192)
95. Linde AD, Linde DA, Mezhlumian A (1995) Do we live in the center of the world? Phys Lett B345:203–210. arXiv:hep-th/9411111 [hep-th] (cit on p 192)
96. Guth AH (2007) Eternal inflation and its implications. J Phys A40 :6811–6826. arXiv:hep-th/0702178 [HEP-TH] (cit on p 192)
97. Steinhardt PJ (1982) Natural inflation. Nuffield workshop on the very early universe, pp 251–266 (cit on p 192)
98. Vilenkin A (1983) The Birth of inflationary universes. Phys Rev D 27:2848 (cit on p 192)
99. Martin J, Ringeval C, Vennin V (2014) Encyclopædia Inflationaris. Phys Dark Univ 5–6:75–235. arXiv:1303.3787 [astro-ph.CO] (cit on p 193)
100. Wald RM (1983) Asymptotic behavior of homogeneous cosmological models in the presence of a positive cosmological constant. Phys Rev D 28:2118–2120 (cit on p 193)
101. Peebles PJE, Vilenkin A (1999) Quintessential inflation. Phys Rev D 59:063505. arXiv:astro-ph/9810509 (cit on p 195)
102. Dimopoulos K, Owen C (2017) Quintessential inflation with α-attractors. JCAP 06:027. arXiv:1703.00305 [gr-qc] (cit on p 195)
103. Hossain W (2018) Quintessential inflation: a unified scenario of inflation and dark energy'. In: Gwak B, Kang G, Kim C et al (eds) EPJ Web Conference vol168, p 04007. arXiv:1801.03272 [gr-qc] (cit on p 195)
104. Dimopoulos K, Donaldson-Wood L (2019) Warm quintessential inflation. Phys Lett B 796:26–31. arXiv:1906.09648 [gr-qc] (cit on p 195)
105. Kallosh R, Linde A, Roest D (2013) Superconformal inflationary α-attractors. JHEP 11:198. arXiv:1311.0472 [hep-th] (cit on p 195)
106. Kallosh R, Linde A, Roest D (2014) Large field inflation and double α-attractors. JHEP 08:052. arXiv:1405.3646 [hep-th] (cit on p 195)
107. Stoeger WR, Ellis GFR, Kirchner U (2004) Multiverses and cosmology: philosophical issue. arXiv:astro-ph/0407329 [astro-ph] (cit on p 195)
108. Laplace PS (1952) A philosophical essay on probabilities. Dover books on science. Dover Publications (cit on pp 196, 253)
109. Easther R, Price LC (2013) Initial conditions and sampling for multifield inflation. JCAP 1307:027. arXiv:1304.4244 [astro-ph.CO] (cit on p 196)
110. Chowdhury D, Martin J, Ringeval C, Vennin V (201) Assessing the scientific status of inflation after Planck. Phys Rev D 100.8, 083537. arXiv:1902.03951 [astro-ph.CO] (cit on p 196)
111. Gibbons GW, Hawking SW, Stewart JM (1987) A natural measure on the set of all universes. Nucl Phys. B281:736 (cit on pp 197, 198, 250, 251)
112. Liouville J (1838) Note sur la Théorie de la Variation des constantes arbitraires. J de Math 3:342 (cit on pp 197, 198)
113. Darboux G (1882) Sur le problème de Pfaff. fr In: Bulletin des sciences Mathématiques et Astronomiques 2e série, 6.1, pp 14–36 (cit on p 198)

114. Gibbs JW (1885) On the fundamental formula of statistical mechanics, with applications to astronomy and thermodynamics. Salem Press (cit on p 198)
115. Gibbons GW, Turok N (2008) The measure problem in Cosmology. Phys Rev D77:063516. arXiv:hep-th/0609095 [hep-th] (cit on pp 199, 250, 251)
116. Li M, Wang Y (2007) The measure for the multiverse and the probability for inflation. JCAP 0706:012. arXiv: 0704.1026 [hep-th] (cit on pp 199, 250)
117. Hawking SW, Don N (1988) How probable is inflation? Nucl Phys B298:789–809 (cit on p 199)
118. Page DN (1987) Probability of R**2 inflation. Phys Rev D36:1607 (cit on p 199)
119. Chmielowski P, Page DN (1988) Probability of Bianchi Type I inflation. Phys Rev D38:2392–2398 (cit on p 199)
120. Remmen GN, Carroll SM (2013) Attractor solutions in scalar-field cosmology. Phys Rev D88:083518. arXiv:1309.2611 [gr-qc] (cit on p 199)
121. Corichi A, Sloan D (2014) Inflationary attractors and their measures. Class Quant Grav 31:062001. arXiv:1310.6399 [gr-qc] (cit on p 199)
122. Schiffrin JS, Wald RM (2012) Measure and probability in Cosmology. Phys Rev D86:023521. arXiv:1202.1818 [gr-qc] (cit on p 200)
123. Eckhardt W (2006) Causal time asymmetry. In: Studies in History and Philosophy of science Part B: studies in history and philosophy of modern physics 37.3. The arrows of time, pp 439–466 (cit on p 200)

Chapter 7
Geometric Initial Conditions for Inflation

I think the Universe is pure geometry – basically, a beautiful shape twisting around and dancing over space-time.
—Antony Garrett Lisi

We have seen the need for a natural, finite, reparametrisation invariant measure of the initial conditions of inflation. In this chapter we shall construct such a measure using the geometric techniques built up throughout this thesis. We shall not consider the problem of initial inhomogeneities, since these have been thoroughly investigated by the works discussed in Sect. 6.4.1, but shall instead focus on the initial field values. This allows us to consider a patch of the Universe that is homogeneous, not only greatly simplifying our equations but also strengthening the link between the physics of the field theory and the geometry of the field space.

7.1 The Phase Space Manifold

Before we specialise to the theory of inflation, let us consider a general theory of N homogeneous scalar fields ϕ^A (collectively ϕ) with a Lagrangian

$$L = \frac{1}{2}k_{AB}(\phi)\dot{\phi}^A\dot{\phi}^B - V(\phi). \tag{7.1}$$

We have seen how we can define an N-dimensional field-space manifold for such a theory and equip it with a metric $G_{AB} = k_{AB}$.

The initial values of the fields ϕ^A can, therefore, by described geometrically as a point on the field space manifold. This geometric description immediately provides us with a reparametrisation invariant volume element

K. Finn, *Geometric Approaches to Quantum Field Theory*, Springer Theses,
https://doi.org/10.1007/978-3-030-85269-6_7

$$dV = \sqrt{G}\, d^N\phi, \tag{7.2}$$

which we can use as a natural measure on the space of these initial field values.

However, the equations of motion for (7.1) are of second order and so choosing an initial value for each field is not enough to fix the evolution of the system. In order to completely specify the initial conditions we must additionally choose an initial value for the derivative of each field $\dot{\phi}^A$. Therefore, we must extend the geometric description (7.1) so that it includes both ϕ^A and $\dot{\phi}^A$.

This motivates us to define the *phase space manifold*, which is a $2N$ dimensional manifold with coordinates

$$\Phi^\alpha = \{\phi^A, \dot{\phi}^A\}. \tag{7.3}$$

We shall use Greek letters to denote indices on this manifold with $\alpha \in \{1, \ldots, 2N\}$. We note that $\dot{\phi}^A$ is an element of the tangent space of the field space manifold. Thus, the phase space manifold is equivalent to the tangent bundle of the field space manifold.

We wish to equip the phase space manifold with a metric so that we can use the techniques of Riemannian geometry. Such a metric should satisfy the following properties:

1. The line element on the phase space manifold should be reparametrisation invariant.
2. The induced metric on the submanifold defined by $\dot{\phi}^A = 0$ should reduce to the field space metric G_{AB}.
3. The induced metric on the submanifold defined by $\phi^A = \phi^A_* =$constant (i.e. the tangent space at the point ϕ^A_*) should be flat for any point ϕ^A_* on the field space manifold.

In order to define such a metric, it is important to note that the quantity $d\dot{\phi}^A$ is not a field space vector. Therefore, we cannot use it to construct the line element of the phase space in a way that satisfies property 1. Instead we utilise the covariant variation

$$D\dot{\phi}^A = d\dot{\phi}^A + \Gamma^A_{BC}\dot{\phi}^B d\phi^C, \tag{7.4}$$

which *is* a field space vector. Here, Γ^A_{BC} are the field space connections.

Armed with this covariant variation we can now define the line element for the phase space manifold as

$$ds^2 = G_{AB} d\phi^A d\phi^B + G_{AB} D\dot{\phi}^A D\dot{\phi}^B. \tag{7.5}$$

We, therefore, see that the phase space metric is given by

$$\mathcal{G}_{\alpha\beta} = \begin{pmatrix} G_{AB} + G_{CD}\Gamma^C_{AE}\Gamma^D_{BF}\dot{\phi}^E\dot{\phi}^F & G_{CB}\Gamma^C_{AD}\dot{\phi}^D \\ G_{AC}\Gamma^C_{DB}\dot{\phi}^D & G_{AB} \end{pmatrix}. \tag{7.6}$$

The metric (7.6) is known as the Sasaki metric on the tangent bundle and was first discovered in [10]. It is the unique[1] metric that satisfies properties 1–3 and is, therefore, the most natural metric for the phase space manifold.

Let us see explicitly how the Sasaki metric satisfies the three properties above. First, since we have only used field space tensors and contracted all field space indices, the line element (7.5) is reparametrisation invariant by construction, in agreement with property 1.

Next, we can calculate the induced metric on the submanifold $\dot{\phi}^A = 0$. This is given by

$$\mathcal{G}_{AB}(\phi, \dot{\phi} = 0) = \mathcal{G}_{\alpha\beta} \frac{\partial \Phi^\alpha}{\partial \phi^A} \frac{\partial \Phi^\beta}{\partial \phi^B}\bigg|_{\dot{\phi}^A = 0} = G_{AB}(\phi) \tag{7.7}$$

in agreement with property 2.

Finally, we see that the induced metric on the tangent space at ϕ^A_* is

$$\mathcal{G}_{AB}(\phi = \phi_*, \dot{\phi}) = \mathcal{G}_{\alpha\beta} \frac{\partial \Phi^\alpha}{\partial \dot{\phi}^A} \frac{\partial \Phi^\beta}{\partial \dot{\phi}^B}\bigg|_{\phi^A = \phi^A_*} = G_{AB}(\phi_*). \tag{7.8}$$

Since $G_{AB}(\phi_*)$ has no dependence on the field derivatives $\dot{\phi}^A$ (which are the coordinates of the tangent space) this is indeed a flat metric in agreement with property 3.

Now that we have a manifold that covers both the field values and their derivatives we can use the invariant volume element of that manifold as a natural measure on the initial conditions of the theory (7.1). We find

$$d\Omega = \sqrt{\det(\mathcal{G}_{\alpha\beta})}\, d^{2n}\Phi \;=\; \det(G_{AB}) d^n\dot{\phi}\, d^n\phi. \tag{7.9}$$

Let us compare this measure to the Liouville measure (6.62) that we discussed in Chap. 6. For the system with Lagrangian (7.1) we find that the canonical variables are

$$q^A = \phi^A, \qquad p_A = G_{AB}\dot{\phi}^B. \tag{7.10}$$

Therefore, the symplectic form (6.61) is

$$\omega = G_{AB} d\dot{\phi}^A \wedge d\phi^B + G_{AB,C}\, \dot{\phi}^A d\phi^C \wedge d\phi^B. \tag{7.11}$$

Plugging (7.11) into (6.62) we see that the only non-vanishing term is

$$d\Omega_L = G_{1A} G_{2B} ... G_{nZ} d\dot{\phi}^A \wedge d\dot{\phi}^B ... \wedge d\dot{\phi}^Z \wedge d\phi^1 \wedge d\phi^2 ... \wedge d\phi^n. \tag{7.12}$$

We can now use the identity

[1] Note that it is only unique up to a relative, dimensionful constant between the two terms in (7.5). Because this constant is dimensionful we can always set it equal to 1 with an appropriate choice of units. The value of this constant will, therefore, not affect the results of this chapter.

$$d\dot{\phi}^A \wedge d\dot{\phi}^B ... \wedge d\dot{\phi}^Z = \epsilon^{AB...Z} d\dot{\phi}^1 \wedge d\dot{\phi}^2 ... \wedge d\dot{\phi}^n, \tag{7.13}$$

to rewrite (7.12) as

$$d\Omega_L = \epsilon^{AB...Z}\, G_{1A} G_{2B} ... G_{nZ}\, d^n\phi d^n\dot{\phi} = \det G\, d^n\phi\, d^n\dot{\phi} = d\Omega. \tag{7.14}$$

We, therefore, see that the measure constructed in this section by geometrical considerations is identical to the Liouville measure. This section, therefore, acts as another justification for using the Liouville measure to study the problem of initial conditions. However, the construction presented here offers a useful geometric interpretation of the Liouville measure and will allow us to go beyond previous work in the literature.

7.2 The Phase Space Manifold for Inflation

Let us now specialise to the case of interest – the theory of inflation. We will consider the Universe to be described by Einstein gravity and dominated by a single scalar field ϕ, which will take on the role of the inflaton. The action for this theory is, therefore, given by

$$S = \int d^4x \sqrt{-g} \left(-\frac{R}{2} + \frac{1}{2}(\partial_\mu \phi)(\partial^\mu \phi) - V(\phi) \right), \tag{7.15}$$

where R is the Ricci scalar, $V(\phi)$ is the inflationary potential, and Lorentz indices are raised and lowered with the help of the spacetime metric $g_{\mu\nu}$ with determinant g.

Note that we will be treating the metric $g_{\mu\nu}$ as a dynamical field and will, therefore, have to deal with the subtleties discussed in Chap. 3. In particular we have seen that specifying the classical action is not enough to fix the physics of this theory and we must additionally choose the value of $\ell(\phi)$. For the discussion in this chapter we shall choose $\ell(\phi) = 1$ for simplicity. We shall consider the effects of relaxing this assumption in Chap. 8.

As we have discussed previously, we are not interested in the effect of inhomogeneities, which have been well studied in the literature (see Sect. 6.4.1). We shall, therefore, focus on a homogeneous patch of the Universe where the inflaton field ϕ has no spatial dependence and spacetime is described by the FRW metric (6.1).

In addition, we shall make the simplifying assumption that the curvature of the Universe is $k = 0$. While this assumption restricts the applicability of our results, we note that inflation is easier to achieve when $k = 0$ than when $k \neq 0$. Indeed, as we have discussed previously, the slow roll regime is a universal attractor when $k = 0$ and all trajectories will eventually inflate as $t \to \infty$, which is not the case when $k \neq 0$. Therefore, the results obtained using the measure developed in this chapter should be considered upper bounds on the likelihood of inflation.

With these assumptions, the spacetime line element can be written in Cartesian coordinates as

$$ds^2 = dt^2 - a(t)^2(dx^2 + dy^2 + dz^2). \tag{7.16}$$

We can calculate the Ricci scalar for this metric and, after some algebra, we find [5]

$$R = -6\left(\left(\frac{\dot{a}}{a}\right)^2 + \frac{\ddot{a}}{a}\right), \tag{7.17}$$

where a dot $\dot{}$ denotes differentiation with respect to time. We also see that the space-time volume element is

$$\sqrt{-g}d^4x = a^3 dt\, dx\, dy\, dz. \tag{7.18}$$

Putting all this together, the action (7.15) is given by

$$S = V_3 \int \left(-3a\dot{a}^2 + \frac{1}{2}a^3\dot{\phi} - a^3 V(\phi)\right) dt, \tag{7.19}$$

where $V_3 = \int d^3x$ is the (infinite) volume of 3-space and we have used integration by parts to remove second derivative terms. The constant factor V_3 is irrelevant to the dynamics and so we can ignore it, setting

$$S = \int L\, dt, \tag{7.20}$$

with the Lagrangian L being given by

$$L = -3a\dot{a}^2 + \frac{1}{2}a^3\dot{\phi} - a^3 V(\phi). \tag{7.21}$$

Notice that if we treat the scale factor a as a scalar field, this Lagrangian is exactly of the form (7.1) and so we can use the methods of the previous section to construct the phase space manifold.

Before we proceed, however, we recall that in Sect. 7.1 we treated the potential term as an external force and it did not feature in the construction of the phase space manifold. But the potential term is incredibly important to the dynamics of inflation and it seems unthinkable that it would not also be important to the selection of initial conditions. In order to be consistent we should include the potential term in the construction of the phase space manifold of inflation in a geometric way.

Fortunately, we already know how to achieve this. As discussed in Chap. 5, we can use the Eisenhart lift to transform the potential term into the kinetic term of a new fictitious field, which can then be interpreted geometrically. We, therefore, add a new scalar field χ to the theory and take the Lagrangian to be

$$L = -3a\dot{a}^2 + \frac{1}{2}a^3\dot{\phi} + \frac{1}{2}\frac{1}{a^3 V(\phi)}\dot{\chi}^2. \tag{7.22}$$

We see that this Lagrangian is still of the form (7.1) and has a three-dimensional field space with a metric

$$G_{AB} = \begin{pmatrix} -6a & 0 & 0 \\ 0 & a^3 & 0 \\ 0 & 0 & \dfrac{1}{a^3 V(\phi)} \end{pmatrix}. \tag{7.23}$$

Let us recall the significance of this field space. The non-zero field-space connections are

$$\begin{aligned} &\Gamma^a_{aa} = \frac{1}{2a}, && \Gamma^a_{\phi\phi} = \frac{a}{4}, && \Gamma^a_{\chi\chi} = -\frac{1}{4a^5 V}, \\ &\Gamma^\phi_{a\phi} = \frac{3}{2a}, && \Gamma^\phi_{\chi\chi} = \frac{V'}{2a^6 V^2}, && \\ &\Gamma^\chi_{a\chi} = \Gamma^\chi_{\chi a} = -\frac{3}{2a}, && && \Gamma^\chi_{\phi\chi} = \Gamma^\chi_{\chi\phi} = -\frac{V'}{2V}. \end{aligned} \tag{7.24}$$

The geodesic equation on the field space manifold is, therefore,

$$\ddot{a} + \frac{1}{2a}\dot{a}^2 + \frac{a}{4}\dot{\phi}^2 - \frac{1}{4a^5 V}\dot{\chi}^2 = 0, \tag{7.25}$$

$$\ddot{\phi} + 3\frac{\dot{a}}{a}\dot{\phi} + \frac{V'}{2a^6 V^2}\dot{\chi}^2 = 0, \tag{7.26}$$

$$\ddot{\chi} - 3\frac{\dot{a}}{a}\dot{\chi} - \frac{V'}{V}\dot{\phi}\dot{\chi} = 0. \tag{7.27}$$

We can rewrite (7.27) as

$$a^3 V \frac{\partial}{\partial t}\left(\frac{\dot{\chi}}{a^3 V}\right) = 0, \tag{7.28}$$

which has a straightforward solution

$$\frac{\dot{\chi}}{a^3 V} = A = \text{constant}. \tag{7.29}$$

Plugging this solution into (7.25) and (7.26) we find, after some rearranging,

$$H^2 + 2\frac{\ddot{a}}{a} = -\frac{1}{2}\dot{\phi}^2 + \frac{A^2}{2}V(\phi), \tag{7.30}$$

$$\ddot{\phi} + 3H\dot{\phi} + \frac{A^2}{2}V'(\phi) = 0, \tag{7.31}$$

where $H \equiv \dot{a}/a$ is the Hubble parameter as before.

We notice that if the constant A satisfies the Eisenhart condition

Table 7.1 The phase space metric for inflation in coordinates $\Phi^\alpha = \{a, \phi, \chi, \dot{a}, \dot{\phi}, \dot{\chi}\}$

$$\mathcal{G}_{\alpha\beta} = \begin{pmatrix} \frac{9\dot{\chi}^2}{4a^5 V} + a\left(\frac{9\dot{\phi}^2}{4} - 6\right) - \frac{3\dot{a}^2}{2a} & \frac{3\dot{\chi}^2 V'}{4a^4 V^2} + \frac{3}{2}a\dot{a}\dot{\phi} & \frac{3\dot{\chi}(a\dot{\phi}V'+2\dot{a}V)}{2a^5 V^2} & -3\dot{a} & \frac{3a^2\dot{\phi}}{2} & -\frac{3\dot{\chi}}{2a^4 V} \\ \frac{3\dot{\chi}^2 V'}{4a^4 V^2} + \frac{3}{2}a\dot{a}\dot{\phi} & \frac{\dot{\chi}^2 (V')^2}{4a^3 V^3} + a^3\left(1 - \frac{3\dot{\phi}^2}{8}\right) + \frac{9\dot{a}^2 a}{4} & \frac{\dot{\chi}\left(3aV^2\dot{\phi}+2a\dot{\phi}(V')^2+12\dot{a}VV'\right)}{8a^4 V^3} & -\frac{3}{2}a^2\dot{\phi} & \frac{3a^2\dot{a}}{2} & -\frac{\dot{\chi}V'}{2a^3 V^2} \\ \frac{3\dot{\chi}(a\dot{\phi}V'+2\dot{a}V)}{2a^5 V^2} & \frac{\dot{\chi}\left(3aV^2\dot{\phi}+2a\dot{\phi}(V')^2+12\dot{a}VV'\right)}{8a^4 V^3} & \frac{2a^6 V\dot{\phi}^2 (V')^2 - 3V^2(\dot{\chi}^2 - 4a^5\dot{a}\dot{\phi}V') + 2(4a^6 + 9\dot{a}^2 a^4)V^3 + 2\dot{\chi}^2 (V')^2}{8a^9 V^4} & \frac{3\dot{\chi}}{2a^4 V} & \frac{\dot{\chi}V'}{2a^3 V^2} & -\frac{a\dot{\phi}V'+3\dot{a}V}{2a^4 V^2} \\ -3\dot{a} & -\frac{3}{2}a^2\dot{\phi} & \frac{3\dot{\chi}}{2a^4 V} & -6a & 0 & 0 \\ \frac{3a^2\dot{\phi}}{2} & \frac{3a^2\dot{a}}{2} & \frac{\dot{\chi}V'}{2a^3 V^2} & 0 & a^3 & 0 \\ -\frac{3\dot{\chi}}{2a^4 V} & -\frac{\dot{\chi}V'}{2a^3 V^2} & -\frac{a\dot{\phi}V'+3\dot{a}V}{2a^4 V^2} & 0 & 0 & \frac{1}{a^3 V} \end{pmatrix}$$

$$A^2 = 2, \tag{7.32}$$

then (7.30) and (7.31) are identical to the Friedmann equations (6.11) and (6.46). Therefore, the dynamics of the early Universe can be described by the geodesic motion of a free particle on this field space manifold.

Now that we have constructed the field space manifold for inflation we can use the techniques described in Sect. 7.1 to construct the corresponding phase space manifold. The phase space will be a 6-dimensional manifold with coordinates

$$\Phi^\alpha = \{a, \phi, \chi, \dot{a}, \dot{\phi}, \dot{\chi}\}. \tag{7.33}$$

The metric of the phase space can be calculated using (7.6) and is shown in Table 7.1.

7.3 Symmetries and Constraints

We cannot immediately use the phase space manifold for inflation, and its associated volume element (7.9), as a measure on the initial conditions of inflation. There are two reasons for this, which we shall address individually.

7.3.1 Symmetries of the Inflationary Phase Space

The first reason that we cannot use (7.9) in its unmodified form is due to symmetries of the equations of motion (7.25)–(7.27). We notice that performing any of the following transformations

$$\chi \to c_1 \chi, \tag{7.34}$$

$$a \to c_2 a, \qquad \chi \to c_2^3 \chi, \qquad \dot{a} \to c_2 \dot{a}, \qquad \dot{\chi} \to c_2^3 \dot{\chi}, \tag{7.35}$$

$$\dot{a} \to c_3 \dot{a}, \qquad \dot{\phi} \to c_3 \dot{\phi}, \qquad \dot{\chi} \to c_3 \dot{\chi}, \tag{7.36}$$

where c_1, c_2 and c_3 are constants, leaves the equations of motion unchanged. Therefore, any two points on the phase space related by one or more of these transformations do not constitute different initial conditions for the Universe since they give rise to an identical evolution. Any such points should, therefore, be identified to avoid double counting.

We can understand these three symmetries physically in the following way. The symmetry (7.34) arises because the fictitious field χ only ever appears with a derivative. Therefore, the absolute value of χ is completely irrelevant.

The second symmetry can be attributed to spatial dilations. Indeed, we see that (7.35) is equivalent to

$$x \to \frac{1}{c_2}x, \qquad y \to \frac{1}{c_2}y, \qquad z \to \frac{1}{c_2}z \tag{7.37}$$

with the transformation of χ following from (7.29). Because we are working in a homogeneous Universe with no characteristic length scales it is clear that the absolute value of the spatial coordinates are unobservable.

Finally, we see that (7.36) is equivalent to a time dilation

$$t \to \frac{1}{c_3}t. \tag{7.38}$$

It may at first seem counterintuitive that such a transformation constitutes a symmetry of our theory, since it is equivalent to the Universe evolving in fast forward or slow motion. However, thanks to the Eisenhart lift, this is indeed the case.

We can understand this symmetry physically with the following argument. As we have seen, the evolution of the Universe can be described by a geodesic on the field space manifold. No amount of time dilation can change that geodesic trajectory, only how quickly it is traversed. Performing the transformation (7.36) simply amounts to changing the affine parametrisation of the geodesic. But an observer in the Universe has no access to the affine parametrisation, only the shape of the trajectory itself. Any measuring device must necessarily be made from the fields themselves and so the observer can only measure the evolution of one field relative to another. In particular, any clock or time-keeping device would be sped up or slowed down by (7.36) at the same rate as all the other fields and so there will be no observable difference in the evolution. If everything in the Universe, including your watch, suddenly started moving twice as fast, would you notice?

We can change coordinates on the phase space manifold in order to isolate the effects of these three symmetry transformations. We, therefore, define

$$\widetilde{\chi} \equiv \frac{\chi}{a^3}, \qquad H_\chi \equiv \frac{\dot{\chi}}{a^3}, \qquad H \equiv \frac{\dot{a}}{a}, \tag{7.39}$$

to isolate the symmetries (7.34) and (7.35) and then define

$$\begin{aligned} H &\equiv \frac{\rho}{\sqrt{6}}\cos\alpha, \\ \dot{\phi} &\equiv \rho\sin\alpha\cos\beta, \\ H_\chi &\equiv \rho\sqrt{V}\sin\alpha\sin\beta. \end{aligned} \tag{7.40}$$

to isolate (7.36).

We can now describe the phase space manifold in terms of the coordinate chart

$$\widetilde{\Phi}^\alpha = \{a, \phi, \widetilde{\chi}, \rho, \alpha, \beta\}. \tag{7.41}$$

Of these coordinates only $\widetilde{\chi}$ is affected by (7.34), only a is affected by (7.35) and only ρ is affected by (7.36). Indeed, we see that, expressed in these new coordinates, the three symmetries become

$$\begin{aligned}\widetilde{\chi} &\to c_1\widetilde{\chi},\\ a &\to c_2 a,\\ \rho &\to c_3\rho.\end{aligned} \tag{7.42}$$

The process of identifying points on the phase space that correspond to the same initial conditions is, therefore, simplified with these coordinates. We simply ignore the values of $\widetilde{\chi}$, a and ρ and, hence, project down to the 3D subspace defined by the coordinates ϕ, α and β.

7.3.2 The Hamiltonian Constraint

The second issue to address is the Hamiltonian constraint, which we discussed briefly in Chap. 6. Recall that, in addition to (6.11), the Einstein equations in an FRW universe also gave the constraint equation (6.10). Our formalism did not reveal this constraint because it arises from the variation of the action with respect to g_{00}, which we have set to 1 with our choice of time coordinate.

We can, therefore, recover this condition by changing time coordinates to $\tau(t)$ so that the line element (7.16) becomes

$$ds^2 = N_L(\tau)^2 d\tau^2 - a(\tau)^2(dx^2 + dy^2 + dz^2), \tag{7.43}$$

where $N_L = dt/d\tau$ is known as the lapse. Following through the algebra, we see that the Lagrangian (7.22), written with this new time coordinate, becomes

$$L = \frac{1}{N_L}\left[-3a\left(\frac{da}{d\tau}\right)^2 + \frac{1}{2}a^3\left(\frac{d\phi}{d\tau}\right)^2 + \frac{1}{2}\frac{1}{a^3V(\phi)}\left(\frac{d\chi}{d\tau}\right)^2\right]. \tag{7.44}$$

We note that we can always recover (7.22) by choosing a time coordinate with $N_L = 1$. However, we must remember that N_L is still a degree of freedom in our theory and, thus, there will still be an equation of motion associated with it. By setting $N_L = 1$ too early, we have missed this equation.

The equation of motion for N_L is precisely the Hamiltonian constraint equation. We can see from (7.44) that it takes the form

$$\mathcal{H} \equiv -3a\dot{a}^2 + \frac{1}{2}a^3\dot{\phi}^2 + \frac{1}{2}\frac{1}{a^3V(\phi)}\dot{\chi}^2 = 0. \tag{7.45}$$

Here, $\mathcal{H}$ is the Hamiltonian of the system, which is equal to the Lagrangian for a purely kinetic theory such as (7.22).

We note that there are no second derivatives in (7.45) and it is, therefore, not an equation of motion, but a constraint on the allowed solutions. Indeed, we see that this condition can be satisfied only on the 5-dimensional Hamiltonian hypersurface, which is a submanifold of the 6-dimensional phase space manifold. Any phase space point that represents a possible initial condition of our Universe must lie on the Hamiltonian hypersurface. In order to construct a measure for the initial conditions of inflation we should, therefore, not consider the full phase space, but only the Hamiltonian hypersurface.

The Hamiltonian hypersurface is still a manifold. It is 5-dimensional and can, therefore, be described by 5 coordinates, which we label φ^i (or collectively by φ). The Hamiltonian hypersurface has a metric induced on it by virtue of its embedding in the phase space manifold. This induced metric is given by

$$\widetilde{\mathcal{G}}_{ij} = \frac{\partial F^\alpha}{\partial \varphi^i}\frac{\partial F^\beta}{\partial \varphi^j}\mathcal{G}_{\alpha\beta}, \tag{7.46}$$

where the functions F^α encode the embedding through

$$\Phi^\alpha = F^\alpha(\varphi). \tag{7.47}$$

The natural measure on the Hamiltonian hypersurface is, therefore, the volume element

$$d\Omega = \sqrt{\det(\widetilde{\mathcal{G}})}\, d^5\varphi. \tag{7.48}$$

In terms of the coordinates (7.41), the Hamiltonian is given by

$$\mathcal{H} = -\frac{1}{2}a^3\rho^2\cos(2\alpha). \tag{7.49}$$

We, therefore, see that the Hamiltonian hypersurface in these coordinates is the hypersurface[2]

$$\alpha = \frac{\pi}{4}. \tag{7.50}$$

We shall describe the Hamiltonian hypersurface with the following coordinates:

$$\varphi^i = \{a, \phi, \widetilde{\chi}, \rho, \beta\}. \tag{7.51}$$

The metric is, therefore, given by

[2] The Hamiltonian constraint is also satisfied for $a = 0$ or $\rho = 0$, but these two conditions yield unacceptable universes with either no space or no time evolution, respectively.

$$\widetilde{\mathcal{G}}_{ij} = \delta_i^\alpha \delta_i^\beta \mathcal{G}_{\alpha\beta}\Big|_{\alpha=\pi/4}, \tag{7.52}$$

where $\mathcal{G}_{\alpha\beta}$ is the metric in Table 7.1 expressed in the coordinates (7.41). This metric is very cumbersome and so we shall not repeat it here.

7.4 Finite Measure for the Initial Conditions of Inflation

There is one final issue with the Hamiltonian hypersurface that we must address before continuing. The metric defined by (7.52) is singular with

$$\widetilde{\mathcal{G}}_{i\rho} = \widetilde{\mathcal{G}}_{\rho i} = 0 \tag{7.53}$$

for all values of the index i. This singularity means that the volume element (7.48) is identically zero and so cannot be used as a measure on the initial conditions.

We can remedy the situation with the following regularisation procedure. Instead of the Hamiltonian hypersurface we consider the surface

$$\mathcal{H} = \epsilon \tag{7.54}$$

for some small constant value ϵ. This surface can be taken arbitrarily close to the Hamiltonian hypersurface by taking the limit $\epsilon \to 0$.

The metric on the hypersurface defined by (7.54) is non-singular and we find that the volume element (7.48) on this surface is given by

$$d\Omega(\epsilon) = \sqrt{\frac{6a^{13}\rho^2\epsilon}{a^3\rho^2 - 2\epsilon}}\,\frac{1}{\sqrt{V(\phi)}}\,d\lambda\, d\phi\, d\widetilde{\chi}\, d\rho\, d\beta. \tag{7.55}$$

As we have previously discussed, the initial values of $\widetilde{\chi}$, a and ρ are irrelevant and should not be considered part of the initial conditions of the Universe. We should, therefore, integrate them out to obtain a measure over the physically relevant degrees of freedom ϕ and β. This gives us

$$d\widetilde{\Omega}(\epsilon) = \int_{\widetilde{\chi}=-\infty}^{\widetilde{\chi}=\infty} \int_{a=0}^{a=\infty} \int_{\rho=0}^{\rho=\infty} d\Omega(\epsilon) = N(\epsilon)\frac{1}{\sqrt{V(\phi)}} d\phi\, d\beta, \tag{7.56}$$

where

$$N(\epsilon) = \int_{-\infty}^{\infty} d\widetilde{\chi} \int_0^{\infty} da \int_0^{\infty} d\rho \sqrt{\frac{6a^{13}\rho^2\epsilon}{a^3\rho^2 - 2\epsilon}} \tag{7.57}$$

is an infinite constant.

Notice how all the dependence on ϵ is now contained in the overall normalisation of the measure $N(\epsilon)$. This normalisation is completely irrelevant as we will only be calculating ratios of the measure. Let us, therefore, normalise the measure so that we have

$$dP = \frac{1}{\int d\widetilde{\Omega}(\epsilon)} d\widetilde{\Omega}(\epsilon) = \frac{1}{2\pi\mathcal{N}} \frac{1}{\sqrt{V(\phi)}} d\phi \, d\beta, \tag{7.58}$$

which we see is independent of ϵ. Here, we have defined

$$\mathcal{N} \equiv \int_{-\infty}^{\infty} \frac{1}{\sqrt{V(\phi)}} d\phi. \tag{7.59}$$

In order for the measure (7.58) to be well defined, we require that the integral (7.59) converges. We see that two conditions are required to achieve this. First, we require that $V(\phi) > 0$ for all values of ϕ. This condition is satisfied for any inflationary potential with a minimum. If there is a minimum in the inflationary potential then it must correspond to the value of the inflaton today, since the inflaton is no longer evolving. But, we can measure the potential energy present in the Universe today and we find that it is positive thanks to the non-zero cosmological constant [6]. Therefore, the minimum of the inflationary potential must be positive and, thus, the condition is satisfied.

The second condition we require is that $V(\phi) \to \infty$ faster than ϕ^2 as $\phi \to \pm\infty$. This is satisfied for some inflationary models, but not all. For example, the integral (7.59) will diverge if the potential has an infinite plateau. For such potentials the measure (7.58) will still require regulating and we will have to work hard to avoid the ambiguities discussed in Sect. 6.4.2. We shall discuss this further in Chap. 8.

However, there is still a wide range of inflationary models for which the integral (7.59) converges, including the simple monomial model $\lambda\phi^4 + \Lambda$. For such models the measure (7.58) is well defined and requires no regularization. This is an incredibly important feature. As we saw in Sect. 6.4.2, much of the confusion over the initial conditions of inflation stems from the infinite nature of the GHS measure. Indeed, we saw that GHS and GT arrived at entirely contradicting results from the same measure simply because they picked different regularisation schemes. The measure defined in (7.58) does not need to be regularized and so will not suffer from these ambiguities.

What does the measure (7.58) say about the likelihood for inflation? We recall that any initial field momentum will be quickly removed by Hubble friction. Therefore, the initial value of $\dot{\phi}$, and hence β will have very little impact on the amount of inflation.

As for the initial value of ϕ, in order to have sufficient inflation to explain the classic cosmological puzzles we require that the inflaton field start high up on its potential so that it has plenty of space to roll down. Specifically, we require the Universe begin with $\phi \gtrsim \phi_c$, where ϕ_c is the critical field value discussed in Chap. 6. However, the measure (7.58) is inversely proportional to the height of the potential and so favours an initial field value close to the minimum. We, therefore, conclude that, generically, initial conditions that allow for inflation are disfavoured by the measure (7.58).

7.5 Effects of Anisotropy

So far we have assumed the Universe is completely isotropic in addition to being homogeneous. Although we observe isotropy today, there is no reason to assume the same degree of isotropy during the early Universe. Indeed, one of the key features of inflation is its ability to smooth out any initial anisotropies [11–14]. We shall, therefore, relax this assumption and consider a homogeneous, but anisotropic universe. Such a universe can be described by a Bianchi I metric [15]

$$ds^2 = dt^2 - a_x^2(t)dx^2 - a_y^2(t)dy^2 - a_z^2(t)dz^2, \tag{7.60}$$

where a_x, a_y and a_z are the scale factors in the x, y and z direction, respectively.

Plugging this metric into (7.15) we find that inflation in a Bianchi I universe is described by the Lagrangian

$$L = -a_x\dot{a}_y\dot{a}_z - a_y\dot{a}_x\dot{a}_z - a_z\dot{a}_x\dot{a}_y + \frac{1}{2}a_xa_ya_z\dot{\phi}^2 - a_xa_ya_zV(\phi). \tag{7.61}$$

In order to describe this theory geometrically, we shall again use the Eisenhart lift. We, therefore, introduce a new scalar field χ and consider the classically equivalent Lagrangian

$$L = -a_x\dot{a}_y\dot{a}_z - a_y\dot{a}_x\dot{a}_z - a_z\dot{a}_x\dot{a}_y + \frac{1}{2}a_xa_ya_z\dot{\phi}^2 + \frac{1}{2}\frac{1}{a_xa_ya_zV(\phi)}\dot{\chi}^2. \tag{7.62}$$

As discussed in Chap. 5, the dynamics of the theory (7.62) can be described as geodesic motion on the field space manifold. We see that the field space manifold in this case is 5-dimensional and can be described by the coordinates $\phi^A = \{a_x, a_y, a_z, \phi, \chi\}$. The metric of the field space can be calculated from (2.41) and is found to be

$$G_{AB} = \begin{pmatrix} 0 & -a_z & -a_y & 0 & 0 \\ -a_z & 0 & -a_x & 0 & 0 \\ -a_y & -a_x & 0 & 0 & 0 \\ 0 & 0 & 0 & a_xa_ya_z & 0 \\ 0 & 0 & 0 & 0 & \frac{1}{a_xa_ya_zV(\phi)} \end{pmatrix}. \tag{7.63}$$

We can now construct the phase space manifold as outlined in Sect. 7.1. In this case the phase space is a 10-dimensional manifold with coordinates

$$\Phi^\alpha = \{a_x, a_y, a_z, \phi, \chi, \dot{a}_x, \dot{a}_y, \dot{a}_z, \dot{\phi}, \dot{\chi}, \}. \tag{7.64}$$

We can calculate the metric on the phase space manifold using (7.6). The form of this metric is extremely cumbersome and not very enlightening and so we shall not display it here.

As before, we must also deal with the symmetries of the theory (7.62) in order to avoid double counting the initial conditions. There are five such symmetries. These are: shifts of χ

$$\chi \to c_\chi \chi, \tag{7.65}$$

three spatial dilations,

$$a_i \to c_i a_i, \qquad \dot{a}_i \to c_i \dot{a}_i, \qquad \chi \to c_i \chi, \qquad \dot{\chi} \to c_i \dot{\chi}, \tag{7.66}$$

for $i \in \{x, y, z\}$, and time dilation,

$$\dot{a}_i \to c_t \dot{a}_i \ \ \forall i, \qquad \dot{\chi} \to c_t \dot{\chi}, \qquad \dot{\phi} \to c_t \dot{\phi}. \tag{7.67}$$

Here, c_χ, c_x, c_y, c_z and c_t are all arbitrary constants.

Let us define a new set of coordinates on the phase space manifold in order to isolate these symmetries. We begin by defining

$$H_i \equiv \frac{\dot{a}_i}{a_i}, \qquad H_\chi \equiv \frac{\dot{\chi}}{a_x a_y a_z}, \qquad \tilde{\chi} \equiv \frac{\chi}{a_x a_y a_z}, \tag{7.68}$$

which isolates the symmetries (7.65) and (7.66).

Before proceeding to isolate the symmetry (7.67), let us define

$$\begin{aligned} H_1 &\equiv \frac{1}{3}\left(H_x + H_y + H_z\right), \\ H_2 &\equiv \frac{1}{6}\left(2H_x - H_y - H_z\right), \\ H_3 &\equiv \frac{1}{\sqrt{12}}\left(H_y - H_z\right). \end{aligned} \tag{7.69}$$

These definitions will simplify the algebra by diagonalising the momentum sector of the phase space metric. This simplification also makes it easier to make connections to our previous work since the isotropic case now corresponds to $H_1 = H$ and $H_2 = H_3 = 0$ and so is easier to identify.

Finally, we define

$$\begin{aligned} H_1 &= \frac{1}{\sqrt{6}}\rho \cos\alpha \cos\gamma, \qquad H_2 = \frac{1}{\sqrt{6}}\rho \cos\alpha \sin\gamma \cos\delta, \\ H_3 &= \frac{1}{\sqrt{6}}\rho \cos\alpha \sin\gamma \sin\delta, \\ H_\chi &= \rho\sqrt{V} \sin\alpha \sin\beta, \qquad \dot{\phi} = \rho \sin\alpha \cos\beta, \end{aligned} \tag{7.70}$$

which isolates the symmetry (7.67). Note that the angles defined above cover the ranges

$$\begin{aligned} \alpha &\in [-\pi/2, \pi/2]\,, & \beta &\in [0, 2\pi]\,, \\ \gamma &\in [0, \pi/2]\,, & \delta &\in [0, 2\pi]\,. \end{aligned} \tag{7.71}$$

In terms of the coordinates

$$\widetilde{\Phi}^\alpha = \{a_x, a_y, a_z, \phi, \widetilde{\chi}, \rho, \alpha, \beta, \gamma, \delta\} \tag{7.72}$$

we see that the five symmetries (7.65)–(7.67) become

$$\begin{aligned} \widetilde{\chi} &\to c_\chi \widetilde{\chi}, \\ a_i &\to c_i a_i\,, \\ \rho &\to c_t \rho. \end{aligned} \tag{7.73}$$

We have, therefore, succeeded in isolating the symmetries and, thus, can avoid double counting by simply ignoring the values of $\widetilde{\chi}$, a_x, a_y, a_z and ρ. This leaves the initial conditions given by ϕ, α and β, which describe the initial field values and derivatives, as before, as well as the new variables γ and δ, which describe the strength and direction of the anisotropy, respectively.

In addition to the discussion of symmetries, we must also consider the Hamiltonian constraint. As before, the Lagrangian (7.62) is purely kinetic and so is equal to the Hamiltonian. The Hamiltonian constraint for this theory is therefore

$$\mathcal{H} = -a_x \dot{a}_y \dot{a}_z - a_y \dot{a}_x \dot{a}_z - a_z \dot{a}_x \dot{a}_y + \frac{1}{2} a_x a_y a_z \dot{\phi}^2 + \frac{1}{2} \frac{1}{a_x a_y a_z V(\phi)} \dot{\chi}^2 = 0. \tag{7.74}$$

We can rewrite the Hamiltonian in terms of the variables (7.72) and we find

$$\mathcal{H} = \frac{1}{2} a_x a_y a_z \rho^2 \left[-\cos^2\alpha \cos(2\gamma) + \sin^2\alpha \right]. \tag{7.75}$$

We, therefore, see that the Hamiltonian hypersurface is defined by the condition

$$\cos(2\gamma) = \tan^2\alpha. \tag{7.76}$$

An important consequence of the constraint (7.76) is that it can only be satisfied if $\cos(2\gamma) \geq 0$. Therefore, we see that the strength of the anisotropy is restricted to be

$$\gamma \leq \frac{\pi}{4}. \tag{7.77}$$

Provided this condition is satisfied we can use (7.76) to eliminate α and, thus, describe the Hamiltonian hypersurface using the coordinates

$$\varphi^i = \{a_x, a_y, a_z, \phi, \widetilde{\chi}, \rho, \beta, \gamma, \delta\}. \tag{7.78}$$

We can calculate the metric on the Hamiltonian hypersurface using (7.46) with the embedding functions given by

$$F^\alpha = \{a_x, a_y, a_z, \phi, \widetilde{\chi}, \rho, \arctan(\sqrt{\cos(2\gamma)}), \beta, \gamma, \delta\}. \tag{7.79}$$

However, as before we find this metric to be singular with

$$\widetilde{\mathcal{G}}_{i\rho} = \widetilde{\mathcal{G}}_{\rho i} = 0 \tag{7.80}$$

for all values of the index i. The volume element on this hypersurface is, therefore, identically zero.

This means we must regularize by considering a surface arbitrarily close to the Hamiltonian hypersurface satisfying (7.54) as we did before. Let us denote the induced metric on this surface by $\widetilde{\mathcal{G}}_{ij}(\epsilon)$. Then the invariant volume element is

$$d\widetilde{\Omega}(\epsilon) = \sqrt{\det \widetilde{\mathcal{G}}_{ij}(\epsilon) d^9\widetilde{\Phi}} \approx \sqrt{\epsilon} \sqrt{\operatorname{adj}[\widetilde{\mathcal{G}}(0)]^{ij} \left. \frac{d\widetilde{\mathcal{G}}_{ij}}{d\epsilon} \right|_{\epsilon=0}} d^9\widetilde{\Phi}, \tag{7.81}$$

where $\operatorname{adj}[\widetilde{\mathcal{G}}]^{ij}$ is the adjugate of $\widetilde{\mathcal{G}}^{ij}$ and we have used Jacobi's identity [16] to evaluate the derivative of the determinant.

We calculate this volume element, taking the limit $\epsilon \to 0$, and find

$$\lim_{\epsilon \to 0} d\widetilde{\Omega}(\epsilon) = \sqrt{\frac{\epsilon}{2} a_x^3 a_y^3 a_z^3 \rho^2} \frac{\sin(\gamma)}{\cos^3(\gamma)} \frac{1}{\sqrt{V(\phi)}} d^9\widetilde{\Phi}. \tag{7.82}$$

We see that the volume element is proportional $\sqrt{\epsilon}$ and so vanishes as $\epsilon \to 0$. But this overall factor will cancel out once the measure is properly normalised and so the limit can be taken safely and our results will not depend on the choice of ϵ.

We can construct a measure on the initial conditions of inflation in a Bianchi I universe by taking this volume element and integrating out the irrelevant degrees of freedom $\widetilde{\chi}$, a_x, a_y, a_z and ρ. We, therefore, obtain

$$\begin{aligned} dP &= \lim_{\epsilon \to 0} \frac{\int_{\widetilde{\chi}=-\infty}^{\widetilde{\chi}=\infty} \int_{a_x=0}^{a_x=\infty} \int_{a_y=0}^{a_y=\infty} \int_{a_z=0}^{a_z=\infty} \int_{\rho=0}^{\rho=\infty} d\Omega(\epsilon)}{\int d\Omega(\epsilon)} \\ &= \frac{1}{2\pi^2 \mathcal{N}} \frac{\sin(\gamma)}{\cos^3(\gamma)} \frac{1}{\sqrt{V(\phi)}} d\phi \, d\beta \, d\gamma \, d\delta, \end{aligned} \tag{7.83}$$

where $\mathcal{N}$ is defined in (7.59).

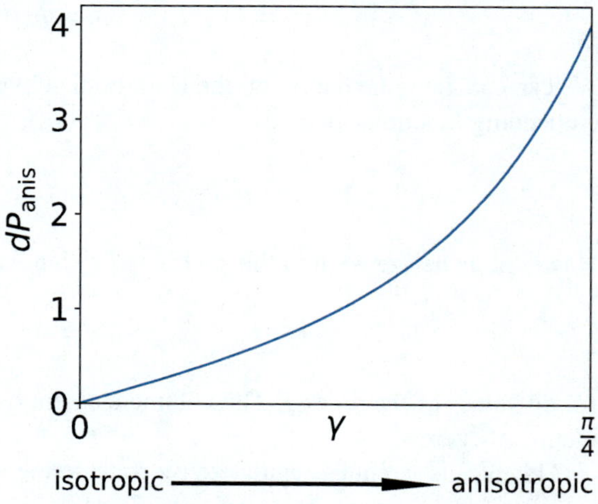

Fig. 7.1 Measure on the initial anisotropy of the Universe as given by (7.85). Figure reproduced from [17]. © 2020 American Physical Society

We see that, provided the integral (7.59) converges, the measure (7.83) is well defined and finite.[3] Therefore, we can calculate a well defined probability of inflation for the same class of theories as in the isotropic case.

In fact, we can see that the measure (7.83) can be factorised into two parts

$$dP = dP_{\text{FRW}}\, dP_{\text{anis}}. \tag{7.84}$$

Here, dP_{FRW} is the measure in an isotropic universe (7.58), and

$$dP_{\text{anis}} = \frac{1}{\pi}\frac{\sin(\gamma)}{\cos^3(\gamma)} d\gamma\, d\delta \tag{7.85}$$

is the measure on the initial anisotropies. Because of this separation we can analyse the likelihood of initial anisotropies independently of the inflationary model.

We see from (7.85) that the measure is uniform over δ. As expected, there is no preference for the anisotropy to point in a particular direction. We can, therefore, integrate out δ to obtain a distribution over the strength of the anisotropy as described by γ. This distribution is shown in Fig. 7.1.

We see that there is a mild preference for more anisotropy over less. However, if inflation does occur it will dilute any anisotropy by an exponential amount [11–13]. The initial anisotropy will, therefore, only impact the amount of inflation if it is exponentially large. Despite the slight preference for anisotropy, we see that this possibility is still highly unlikely according to the measure (7.85) and we, therefore, conclude that allowing for initial anisotropy will have a negligible effect on our previous results.

[3] Note that the measure is finite in the γ direction because of the restriction (7.77).

7.6 Example: ϕ^4 Inflation

To conclude this chapter, let us consider a specific inflationary model. This will allow us to be concrete and quantitative about the likelihood of inflation according to the measure constructed above.

We shall take as our example the simple model

$$V(\phi) = \lambda\phi^4 + \Lambda. \tag{7.86}$$

Although this model is disfavoured at the 2σ level by the non-observation of tensor modes in the CMB [6] it is still worth analysing as the simplest model for which the normalisation constant (7.59) is finite. The value of the quartic coupling λ can be inferred from measurements of the amplitude of fluctuations in the CMB. Using the latest data from the Planck satellite [9] we find $\lambda \approx 5 \times 10^{-14}$.

We note that the cosmological constant Λ is usually ignored in studies of this potential. This is because the observed value today is incredibly small. Indeed, the Planck satellite measures $\Lambda = 2.846 \times 10^{-122}$ [6] in Planck units. The reason for this tiny value is still an unsolved problem and there is much ongoing work to try to solve it [1–4].

However, for the purposes of studying the measure (7.85) it is *essential* to have a non-zero value of Λ. Ignoring this term would cause the integral (7.59) to blow up at $\phi = 0$ and, hence, would leave the measure (7.85) undefined. Because the value of Λ has a significant effect on our results and its current value is poorly understood theoretically we shall not fix it to the value observed by Planck. Instead we shall allow for the possibility that Λ was much larger at the time of inflation and, therefore, treat it as a free parameter.

We note, however, that we cannot allow for any value of Λ. To begin with, we require that Λ be strictly positive in order for the integral (7.59) to converge. In addition, the slow roll parameter for the potential (7.86) is

$$\epsilon_V = \frac{8\lambda^2\phi^6}{\lambda^2\phi^8 + 2\lambda\Lambda\phi^4 + \Lambda^2} \tag{7.87}$$

and so we see that if Λ is greater than $27\lambda/4$ then $\epsilon_V < 1$ for all ϕ and inflation can never end. We must, therefore, only consider values of the cosmological constant

$$0 < \Lambda < \frac{27}{4}\lambda \approx 3.4 \times 10^{-13}. \tag{7.88}$$

Assuming that these conditions are met, the normalisation constant (7.59) can be calculated to give

$$\mathcal{N} = \frac{8}{\sqrt{\pi}}(\lambda\Lambda)^{-\frac{1}{4}}\Gamma\left(\frac{5}{4}\right)^2, \tag{7.89}$$

where $\Gamma\left(\frac{5}{4}\right) \approx 0.906$ is the Euler Gamma function. We see that $\mathcal{N}$ is finite as promised. Thus, the measure (7.85) for this theory is well defined.

As discovered in the previous section, the distribution of initial anisotropy is model independent and has little impact on our results. We shall, therefore, focus on the measure of initial field values (7.58). For the theory (7.86), this is given by

$$dP = \frac{\sqrt{\pi}}{8\Gamma\left(\frac{5}{4}\right)^2}(\lambda\Lambda)^{\frac{1}{4}}\frac{1}{\sqrt{\lambda\phi^4+\Lambda}}d\phi\, d\beta. \tag{7.90}$$

We can rewrite β in terms of the more physically meaningful quantity $\dot{\phi}$ using the definitions (7.40), the Hamiltonian constraint (7.50) and the Eisenhart condition (7.32). We find

$$\dot{\phi} = \frac{1}{\tan\beta}\sqrt{2V(\phi)}. \tag{7.91}$$

We can, therefore, rewrite (7.90) as

$$dP = \frac{\sqrt{2\pi}}{8\Gamma\left(\frac{5}{4}\right)^2}(\lambda\Lambda)^{\frac{1}{4}}\frac{1}{2\lambda\phi^4+2\Lambda+\dot{\phi}^2}d\phi d\dot{\phi}. \tag{7.92}$$

This measure is shown in Fig. 7.2 for two different values of Λ.

We notice that the measure is tightly concentrated around the origin $\phi = \dot{\phi} = 0$ and that this concentration becomes more pronounced as we reduce the value of Λ. This is entirely expected. We can see from (7.92) that the measure at the origin is proportional to $\Lambda^{-3/4}$ and, thus, diverges as $\Lambda \to 0$.

Let us now analyse what this measure means for the likelihood of inflation in this model. As we have discussed earlier, in order to successfully solve the classic cosmological puzzles we need at least $N > 60$ e-foldings of slow roll inflation. We must, therefore, enter the slow roll regime with $\phi > \phi_c$, where ϕ_c is the critical value discussed in Chap. 6. The critical value ϕ_c can be calculated from (6.59) and we find that it satisfies

$$\int_{\phi_c}^{\phi_f} \frac{V(\phi)}{V'(\phi)}d\phi = 60, \tag{7.93}$$

where ϕ_f is the value of ϕ at the end of inflation, given by

$$\epsilon_V(\phi_f) = 1. \tag{7.94}$$

For the potential (7.86) we find that

$$\phi_f \approx 2.82M_P, \qquad \phi_c \approx 22.09M_P. \tag{7.95}$$

These values are very insensitive to the values of both λ and Λ. They are insensitive to Λ because inflation happens at large values of ϕ, for which the second term in (7.86)

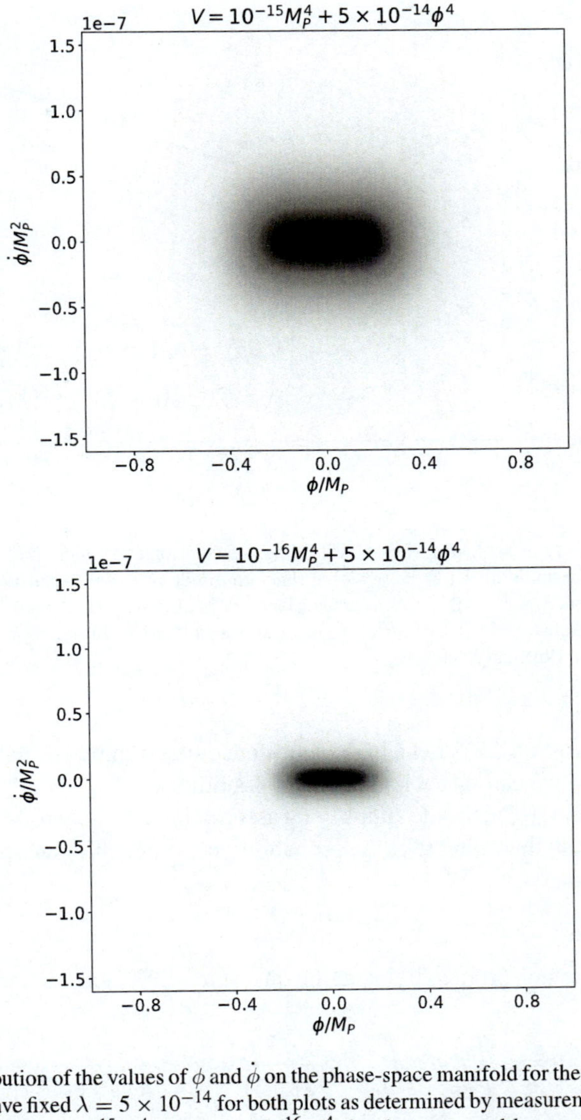

Fig. 7.2 Distribution of the values of ϕ and $\dot{\phi}$ on the phase-space manifold for the potential $V(\phi) = \lambda\phi^4 + \Lambda$. We have fixed $\lambda = 5 \times 10^{-14}$ for both plots as determined by measurements of the CMB, and have chosen $\Lambda = 10^{-15} M_P^4$ and $\Lambda = 10^{-16} M_P^4$ for the upper and lower plots, respectively. Darker areas correspond to a higher value of the measure (7.92). Figure reproduced from [18].

is negligible. Furthermore, when we can ignore Λ, we see that the value of λ cancels out in both (7.87) and (7.93). Therefore, these values are also insensitive to λ.

Because of the expansion of the Universe, any initial field momentum will be very quickly damped by Hubble friction. Therefore, the Universe would have to start with an exponentially high value of $\dot{\phi}$ for this to affect the amount of inflation [7, 8]. As

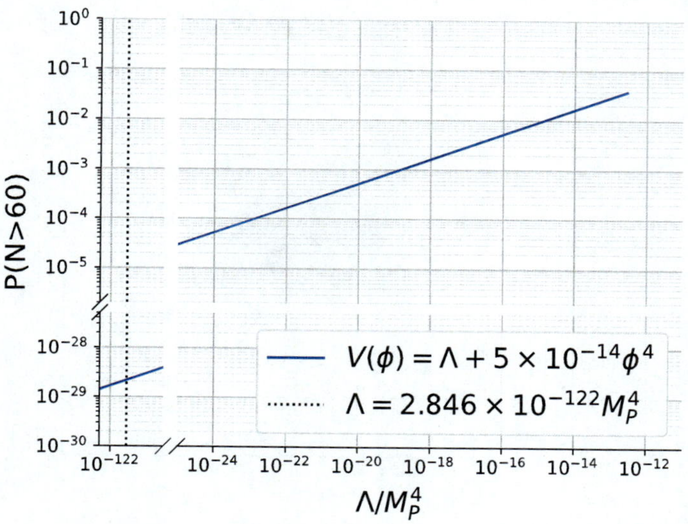

Fig. 7.3 Probability of achieving $N > 60$ e-foldings of inflation as a function of the cosmological constant Λ for the potential $V(\phi) = \lambda\phi^4 + \Lambda$. The amplitude of scalar perturbations in the CMB has been used to set $\lambda = 5 \times 10^{-14}$. A vertical dashed line is placed at $\Lambda = 2.846 \times 10^{-122} M_P^4$ corresponding to the observed value of the cosmological constant today. Figure reproduced from [18].

we can see from Fig. 7.2, such a high value constitutes a minute fraction of the total measure and so we can safely ignore such possibilities.

We can, therefore, make the simplifying assumption that slow roll inflation begins immediately and, therefore, take the probability of sufficient inflation to be

$$P(N > 60) \approx P(|\phi| > \phi_c). \tag{7.96}$$

We can evaluate this probability using the measure (7.92) and we find

$$\begin{aligned} P(N > 60) &= 2\int_{-\infty}^{\infty} d\dot{\phi} \int_{\phi_c}^{\infty} d\phi \, \frac{\sqrt{2\pi}}{8\Gamma\left(\frac{5}{4}\right)^2} (\lambda\Lambda)^{\frac{1}{4}} \frac{1}{2\lambda\phi^4 + 2\Lambda + \dot{\phi}^2} \\ &= \frac{\sqrt{\pi}\, {}_2F_1\left(\frac{1}{4}, \frac{1}{2}; \frac{5}{4}; -\frac{\Lambda}{\lambda\phi_c}\right)}{4\phi_c \Gamma\left(\frac{5}{4}\right)^2 \sqrt[4]{\frac{\lambda}{\Lambda}}}, \end{aligned} \tag{7.97}$$

where ${}_2F_1$ is the hypergeometric function. This distribution is shown in Fig. 7.3 as a function of the cosmological constant Λ.

We see from Fig. 7.3 that the probability of obtaining $N > 60$ e-foldings of inflation is absolutely tiny. Indeed, if we take the current value of the cosmological constant at face value then the measure (7.58) gives a probability of 2.1×10^{-29} for the likelihood of inflation. Even if we consider the possibility that Λ was much bigger in the past, we see that the likelihood of inflation never goes above a few percent.

This means that, even with the most favourable choice of parameters, the most likely scenario is that we live in a Universe without enough inflation to explain the cosmological puzzles. In light of this result, we are faced with an uncomfortable truth; if we cannot design a potential that avoids this issue, we must work hard to explain why we find ourselves in one of the minority of universes where there is enough inflation. Otherwise, we will be forced to abandon inflation as a solution to the problems of standard HBB cosmology.

References

1. Martin J (2012) Everything you always wanted to know about the cosmological constant problem (but were afraid to ask). Comptes Rendus Physique 13:566–665. arXiv: 1205.3365 [astro-ph.CO] (cit. on pp. 162, 233, 249)
2. Sola J (2013) Cosmological constant and vacuum energy: old and new ideas. J Phys Conf Ser 453:012015, Papakostas T, Pliakis DA (eds). arXiv: 1306.1527 [gr-qc] (cit. on pp. 162, 233, 249)
3. Frusciante N, Perenon L (2020) Effective field theory of dark energy: a review. Phys Rept 857:1–63. arXiv: 1907.03150 [astro-ph.CO] (cit. on pp. 162, 233, 249)
4. Carroll SM (2001) The cosmological constant. Living Rev Rel 4:1. arXiv: astro-ph/0004075 (cit. on pp. 162, 233, 249)
5. Gorbunov DS, Rubakov VA (2011) Introduction to the theory of the early universe: hot big bang theory. World Scientific (cit. on pp. 174, 216)
6. Aghanim N et al (2018) Planck 2018 results. VI. Cosmological parameters. arXiv:1807.06209 [astro-ph.CO] (cit. on pp. 179, 180, 226, 233)
7. Salopek DS, Bond JR (1990) Nonlinear evolution of long wavelength metric fluctuations in inflationary models. Phys Rev D42:3936–3962 (cit. on pp. 187, 193, 237)
8. Liddle AR, Parsons P, Barrow JD (1994) Formalizing the slow roll approximation in inflation. Phys Rev D50:7222–7232. arXiv: astro-ph/9408015 [astro-ph] (cit. on pp. 187, 193, 237)
9. Akrami Y et al (2018) Planck 2018 results. X. Constraints on inflation. arXiv: 1807.06211 [astro-ph.CO] (cit. on pp. 190, 233, 252)
10. Sasaki S (1958) On the differential geometry of tangent bundles of Riemannian manifolds. Tohoku Math J (2) 10:338–354. arXiv: 1011.4249 [gr-qc] (cit. on p. 213)
11. Collins CB, Hawking SW (1973) Why is the Universe isotropic? Astrophys J 180:317–334 (cit. on pp. 227, 232)
12. Moss I, Sahni V (1986) Anisotropy in the chaotic inflationary universe. Phys Lett B178:159–162 (cit. on pp. 227, 232)
13. Barrow JD, Gron O (1986) Homogenization and isotropization of an inflationary cosmological model. Phys Lett B182:25–27 (cit. on pp. 227, 232)
14. Jensen LG, Stein-Schabes JA (1986) The effect of inflation on anisotropic cosmologies. Phys Rev D34:931 (cit. on p. 227)
15. Bianchi L (1898) On the three-dimensional spaces which admit a continuous group of motions. Memorie di Matematica e di Fisica della Società Italiana delle Scienze 11:267–352 (cit. on p. 227)
16. Jacobi CGJ (1841) De formatione et proprietatibus Determinantium. J Reine Angew Math 22:285–318 (cit. on p. 231)
17. Kieran Finn. "Initial Conditions of Inflation in a Bianchi I Universe". In: Phys. Rev. D 101.6 (2020), p. 063512. arXiv: 1912.04306 [gr-qc] (cit. on p. 233)
18. Kieran Finn and Sotirios Karamitsos. "Finite measure for the initial conditions of inflation". In: Phys. Rev. D 99.6 (2019). [Erratum: Phys.Rev.D 99, 109901 (2019)], p. 063515. arXiv: 1812.07095 [gr-qc] (cit. on pp. 236, 238)

Chapter 8
Conclusions

> *Equations are just the boring parts of mathematics. I attempt to see things in terms of geometry.*
>
> —Stephen Hawking

Nature doesn't care about our conventions. She doesn't care about our coordinates, our choice of units, or the way we choose to parametrise our fields. The techniques of field space covariance, together with the Vilkovisky DeWitt effective action [1–3] make these facts manifest.

This allows us to draw a clear distinction between two often confused concepts: the *content* of a theory and its *representation*. The content of a theory is the physical predictions it makes. It can be thought of as a dictionary linking every possible experiment with the result of that experiment as predicted by the theory.

The representation of a theory, on the other hand, is the way we have chosen to write the theory down. It is impractical, and not very useful, to simply list the result of each experiment, so theories must be expressed through mathematical objects and a set of rules. For a QFT, the representation is usually given in terms of a set of quantum fields and a classical action as well as the rules of Feynman diagrams and the quantum effective action.

We have seen that the standard way in which Feynman diagrams and the effective action are defined is not reparametrisation invariant. We are, therefore, required to specify some preferred frame in which these definitions hold. The existence of a preferred frame blurs the line between content and representation since the physical predictions of the theory now depend on the parametrisation of the fields in this frame.

In contrast, the covariant techniques developed in this thesis are fully reparametrisation invariant and no parametrisation is more preferred than any other. Thus, the content of a theory can be specified completely independently of its representation.

Indeed, we have seen that a quantum field theory can be described geometrically in terms of a field space manifold. The model functions, which specify the content

K. Finn, *Geometric Approaches to Quantum Field Theory*, Springer Theses,
https://doi.org/10.1007/978-3-030-85269-6_8

of the theory can then be given as tensors on that manifold, which can be defined geometrically without ever referencing a particular parametrisation of the fields.

We have seen that maintaining this separation between the content and representation of a theory becomes more involved when gravity is treated dynamically. In this case we must define the field space manifold carefully in order to maintain both manifest reparametrisation invariance and spacetime diffeomorphism invariance. This leads us to define the metric of spacetime $\bar{g}_{\mu\nu}$ in a reparametrisation invariant way using the model function $\ell(\phi)$. We note that the relationship between the tensor field $g_{\mu\nu}$ and the metric of spacetime $\bar{g}_{\mu\nu}$ cannot be inferred from the classical action and, thus, $\ell(\phi)$ represents an additional piece of information that must be chosen to fully define a scalar-tensor theory of gravity.

The model function $\ell(\phi)$ affects many quantities used in QFT. These include the field space covariant definitions of the Dirac delta function, the functional derivative, the functional determinant and the path integral volume element. However, in many cases the dependence on $\ell(\phi)$ cancels out. For example,

$$X_{\bar{A}} \frac{\bar{\delta} F}{\bar{\delta} \phi^{\bar{A}}} = X_{\widehat{A}} \frac{\delta F}{\delta \phi^{\widehat{A}}} \tag{8.1}$$

for any configuration space vector $X_{\bar{A}}$ and scalar F. In the expression for the VDW effective action (3.63), the only factor of $\ell(\phi)$ that does not cancel is in the definition of the configuration space metric.

Despite this, it is still important to use the covariant definitions developed in Chap. 3. The only way to maintain manifest invariance under both field reparametrisations and spacetime diffeomorphisms is to ensure that the metric of spacetime is a scalar under diffeomorphisms of the field space and vice versa. Without this separation, the formalism will inevitably be either parametrisation dependent (for example if $g_{\mu\nu}$ is used as the spacetime metric) or spacetime diffeomorphism dependent (if the factors of $g_{\mu\nu}$ are removed altogether).

This is especially important when calculating correlation functions using derivatives of the effective action. The ordinary functional derivative $\delta\Gamma/\delta\Phi^{\widehat{I}}$ is not a configuration space vector when gravity is dynamical, even when the effective action Γ is a configuration space scalar.

Therefore, while the effects of $\ell(\phi)$ can be captured by a formalism with a preferred frame such as in [4], such an approach misses important aspects. Without a frame invariant definition of the metric of spacetime we are doomed to fall into the trap of the cosmological frame problem.

We have seen that the standard definition of the quantum effective action is frame dependent, while the VDW effective action is not. They can only agree, therefore, in (at most) one frame. The two expressions will agree if we transform to a frame in which $\ell(\phi) = 1$ in Planck units and where, additionally, all fields are canonically normalised.

Such a frame can only exist if the field space manifold is flat. Thus, for theories with a curved field space the two formalisms do not agree in any frame, and the non-trivial effects of the field space must be taken into account regardless of the parametrisation of the fields.

One such theory is General Relativity, as described by the Einstein Hilbert action. We showed in Chap. 3 that the field space of General Relativity has a negative curvature. Thus, quantum corrections to gravity will always contain non-trivial field-space effects. These effects have so far been ignored in attempts to construct a UV complete theory of quantum gravity. It would be interesting to investigate whether the reason such a construction has proven so difficult is related to this negative curvature.

We have seen that fermions can also be described geometrically. In this case the field space must be generalised to a supermanifold in order to take into account the anticommutativity of the spinor fields.

Thus, the formalism of field space covariance can now be applied to any QFT, regardless of its field content. This means that it should be straightforward to construct the field space manifold for realistic models of particle physics such as the Standard Model.

With the addition of fermions, however, the identification of the metric becomes less straightforward. For bosonic degrees of freedom, the standard kinetic term comes with two derivatives of the fields. These are connected by a rank $\binom{0}{2}$ field space tensor, which can naturally be interpreted as the metric. Fermions, on the other hand, enter the Lagrangian with only a single derivative and thus there is no analogous tensor to identify as the metric.

As we have seen, it is still possible to define a metric for fermionic theories, but the process is more involved. Instead of calculating the metric directly from the action, we are required to solve an implicit equation for the field space vielbeins. These additional steps mean that the relation between the geometry of the field space and the physics of the field theory is less apparent. In particular, since the Lagrangian is not proportional to the field space line element (as it is for bosonic theories), the classical equations of motion of the theory are no longer related to the geodesic equation of the field space. An interesting avenue for future research would be to investigate what links still remain and what physical phenomena can be related to geometric objects on the field space in the presence of fermions.

Nonetheless, the techniques of Vilkovisky and DeWitt apply just as well to fermionic theories as they do to bosonic ones. We are, therefore, able to construct the VDW effective action for such theories. This action is manifestly invariant under reparametrisations of both the scalar and fermion fields.

We have so far only considered field reparametrisations of the form

$$\Phi^\alpha \to \widetilde{\Phi}^\alpha = \widetilde{\Phi}^\alpha(\boldsymbol{\Phi}). \tag{8.2}$$

This covers the vast majority of reparametrisations that we are likely to encounter, but it doesn't cover everything. In particular, field reparametrisations that depend on derivatives of the fields are not covered by (8.2).

One example of such a reparametrisation is

$$\phi \to \tilde{\phi} = \phi + \bar{\epsilon}\psi, \qquad \psi \to \tilde{\psi} = \psi + \gamma^\mu \partial_\mu \phi \, \epsilon, \tag{8.3}$$

which underpins the theory of Supersymmetry [5]. Here, ϕ and ψ are a scalar and fermion field, respectively, and ϵ is an infinitesimal Grassmannian parameter.

In addition, even within pure scalar field theories there are some reparametrisations that we have not considered, for example reparametrisations of the form

$$\phi^A \to \widetilde{\phi}^A = \phi^A + M^A_{BC} g^{\mu\nu} \partial_\mu \phi^B \partial_\nu \phi^C. \tag{8.4}$$

When gravity is treated dynamically, new derivative-dependent reparametrisations become available, including transformations involving the Ricci scalar [6]

$$g_{\mu\nu} \to \widetilde{g}_{\mu\nu} = F(R) g_{\mu\nu} \tag{8.5}$$

and *disformal* transformations [7, 8] of the form

$$g_{\mu\nu} \to \widetilde{g}_{\mu\nu} = g_{\mu\nu} + N_{AB} \nabla_\mu \phi^A \nabla_\nu \phi^B, \tag{8.6}$$

as well as more complex transformations involving second and third derivatives of the fields [9].

It should be straightforward to include such transformations by generalising (8.2) to reparametrisations of the configuration space

$$\Phi^{\widehat{\alpha}} \to \widetilde{\Phi}^{\widehat{\alpha}} = \widetilde{\Phi}^{\widehat{\alpha}}(\mathbf{\Phi}). \tag{8.7}$$

After all, the derivative of a field is simply the difference between the value of the field at two points in spacetime, which are both configuration space coordinates. However, further investigation is required to ensure that the formalism can handle such transformations while maintaining manifest spacetime diffeomorphism invariance.

In this thesis we have assumed that the connection on the field space manifold is the Levi-Civita connection (2.9) obtained from the field space metric. Another possible avenue for future research would be to investigate the consequences of a more general field space connection, perhaps unrelated to the metric. This would allow the possibility of theories with non-zero field space torsion.

We should note, however, that such torsion could only appear in theories with a curved field space. For theories with a flat field space we can always switch to a parametrisation in which all the fields are canonically normalised, at which point the effects of the field space must vanish. Thus, for such theories, the field space connections must be given by the Christoffel symbols. Nonetheless, it would be interesting to investigate the consequences of torsion in theories with curved field spaces.

The construction of the field space manifold is usually performed without ever referencing the potential term. However, we have seen that this term too can be described geometrically. This was achieved using the Eisenhart lift.

The Eisenhart lift, when applied to classical mechanics, can be used to describe any conservative force as the consequence of the curvature of spacetime. Therefore, just as Einstein showed that gravity is a fictitious force and Kaluza and Klein showed the same for electromagnetism, we can now treat any force as fictitious, even, for example, the restoring force of a simple harmonic oscillator.

Applied to QFT, we have seen that the force arising from the potential term can be described by the curvature of field space. By introducing new degrees of freedom we can construct a purely kinetic Lagrangian to describe any QFT. In this way any potential term can be replaced by a kinetic term.

We have, so far, only considered applying the Eisenhart lift to the entire potential. However, we have seen that the method works just as well if applied to an individual term. In fact, by adding *multiple* new vector fields we could even lift several terms independently. This would result in multiple $\hat{\pi}_0$ operators, each with their own eigenvalue. The value of these eigenvalues would dictate the relative strength of each interaction in the QFT.

Although we have, so far, only considered states that are eigenstates of the $\hat{\pi}_0$ operators, the most general state is a superposition of these. Since different eigenstates are orthogonal, we can interpret such a superposition as a multiverse of non-interacting universes. Each of these universes would have different eigenvalues and, hence, a different set of interactions in the QFT.

The existence of such a multiverse may give us a new outlook on fine-tuning problems such as the gauge hierarchy problem [10, 11] and the cosmological constant problem [12–16]. For example, by including a term in the Lagrangian

$$\mathcal{L} \ni \frac{1}{2M^2} \nabla_\mu B^\mu \nabla_\nu B^\nu, \tag{8.8}$$

the size of the cosmological constant would be dictated by the eigenvalue of $\hat{\pi}_0$ and would, hence, differ across the multiverse. As first shown by Weinberg [17],[1] only universes with a small cosmological constant would evolve to form life and, therefore, in such a theory it would be fully expected that we should observe a tiny value today.

One of the most developed applications of field space covariance is to the theory of inflation [20–22]. As we have seen, a period of exponential expansion of the Universe, driven by a slowly rolling scalar field, can account for the flatness, homogeneity and lack of monopoles observed today as well as the spectrum of perturbations observed in the CMB.

At leading order, inflation can be described by a homogeneous scalar field theory in a universe described by the FRW metric. Having investigated such theories in

[1] See also [18, 19] for more recent work on this idea.

Chaps. 2, 6 and 7 we know that the dynamics will be determined by both the curvature of the field space and the inflationary potential.

If there is only a single inflaton field, the field space will necessarily be flat and so curvature effects can be ignored. However, for theories with multiple inflaton fields these effects can be very important. For example, the curvature of field space can destabilise the minimum of the potential, causing the inflaton fields to follow an entirely different trajectory than would be naively expected [23–25]. Field space curvature can also have an impact on the spectrum of produced perturbations, leading to a higher level of non-Gaussianity than would otherwise be expected [26–31]. In fact, the field space effects can be so extreme that the inflaton never enters slow roll at all [32–34].

Instead of ignoring these field space effects, we should embrace them. We have seen how, by using the Eisenhart lift, we can incorporate the inflationary potential into the field space. In this way the dynamics of inflation are described purely geometrically as the geodesic of a manifold.

Without the potential to set the scale we may wonder how to identify the slow roll regime once the theory has been lifted in this way. However, from (5.82), we see that the slow roll regime is determined by

$$\dot{\phi}^2 \ll M^2 \dot{\chi} \tag{8.9}$$

This condition is independent of the choice of the mass scale M since the Eisenhart condition (7.32) requires that $\dot{\chi}$ scale as M^{-2}. The slow roll condition is, therefore, replaced by a slow-roll hierarchy in which the system is evolving faster in the fictitious direction than in any other.

Describing inflation geometrically has another advantage, namely a natural measure on the space of initial conditions. The cosmological puzzles that inflation sets out to solve are fundamentally problems of finely tuned initial conditions. Therefore, it is imperative that we understand whether the initial conditions required to start inflation are finely tuned or not.

We have seen how previous attempts to quantify the fine tuning of inflation have used measures that are infinite in extent [35–37]. They have, therefore, been unable to use Laplace's principle of indifference without first regulating the measure. The ambiguities arising from the choice of regularisation have led to huge disagreement between authors about the likelihood of inflation.

Using the Eisenhart lift and the methods of field space covariance, we were able to overcome these problems and define a natural measure that is finite. By considering the induced metric on the Hamiltonian hypersurface, and integrating out the unphysical degrees of freedom, we were able to define a manifold of initial conditions with a finite total volume. This manifold is shown schematically in Fig. 8.1 for the example potential $V(\phi) = \phi^4 + 0.01 M_P^4$ using an equal-area projection on the cylinder. While the tendrils do stretch to infinity, the total volume of this manifold is finite as shown in Chap. 7.

Fig. 8.1 Equal-area projection of the manifold of initial conditions for inflation. We have used an example potential of $V(\phi) = \phi^4 + 0.01 M_P^4$

The finite nature of the initial conditions manifold allows us to take the following, frequentist, approach to probabilities in inflation. We consider the quantum gravity regime to be a "blind creator" which starts the Universe by "throwing a dart" at the manifold of initial conditions. In the absence of any further information about the early Universe, we take Laplace's principle of indifference and assume that this dart is equally likely to hit any point on the manifold. Thus, the probability of achieving sufficient inflation to explain the cosmological puzzles is equal to the fraction of the manifold that corresponds to acceptable initial conditions.

We note that this approach to probability differs from the works by Gibbons, Hawking and Stewart [35] and Gibbons and Turok [36] discussed in Chap. 6. Instead of counting the different possible trajectories through phase space, we consider all possible starting points for those trajectories. Therefore, our measure is one dimension higher than the corresponding GHS measure since we count different points on an inflationary trajectory as different initial conditions.

The reason for taking this approach is threefold. First, it removes the need for an arbitrary counting surface, which, as we have seen, can influence the results. Second, we believe that if a particular trajectory takes up a larger portion of the measure then it should be considered more likely, something that is lost if we just count trajectories. Finally, this approach avoids the temptation to use Bayesian probability, which, as we have seen, can lead to incorrect results when applied to the Universe as a whole.

We have seen that for simple models of inflation, such as $\lambda\phi^4$, the probability obtained in this manner is very low. Even if we allow the cosmological constant Λ to be much larger at the time of inflation than it is today, the probability is only a few percent. If we instead take the current observations of $\Lambda = 2.846 \times 10^{-122} M_P^4$ at face value, we are forced to conclude that the probability of inflation is incredibly tiny, on the order of 10^{-29}.

It may be that more complex models of inflation do not suffer from this problem. For example, current observations of the CMB favour inflationary theories with a long plateau at energy $V_p \ll M_P$ [38]. If this plateau is long enough, it might be able to overcome the $\sim \sqrt{\Lambda / V_p}$ suppression coming from the phase space measure. However, if the plateau is infinite in extent, as it is in many models, then the manifold of initial conditions will no longer be finite and we will be required to regularise.

One way to regularise would be to insert a wall in the potential at some finite value of ϕ. This may happen naturally, for example from quantum corrections to the classical potential. As we have discussed previously, we expect big effects from quantum gravity when the field space distance from the minimum is greater than the Planck mass [39]. These corrections may be enough to cause the integral (7.59) to converge.

An alternative way to naturally regularise the measure would be to introduce a non-trivial value for the model function $\ell(\phi)$. In Chap. 7 we set $\ell = 1$, but it is relatively straightforward to include its effects. As we can see from (3.33), the field space metric in the presence of a non-trivial ℓ is simply multiplied by a factor ℓ^2. Since the Hamiltonian hypersurface is 5-dimensional, the initial conditions measure will thus receive a prefactor of ℓ^{10}. The total measure will, therefore, converge provided $1/\ell$ grows quicker than $\phi^{\frac{1}{5}}$ as $\phi \to \infty$, even for an infinite plateau.

Once we have a better understanding of the nature of quantum gravity we might hope to be able to calculate directly the expected distribution of field values in the early Universe and it may be that this distribution differs from the one calculated in Chap. 7. However, until we have a UV complete theory, our best approach is to treat quantum gravity as a black box and, hence, rely on Laplace's principle of indifference [40]. After all, without a thorough understanding of quantum gravitational effects we have no reason to assume that they will enhance, rather than diminish, the probability of inflation, especially not by the enormous factor needed to overcome the suppression calculated in Chap. 7.

We have so far only considered inflation in a flat, homogeneous universe. The next logical step would be to relax these assumptions. We have already seen that relaxing the assumption of isotropy does not affect the results in a significant way and so we may hope that allowing for spatial curvature or initial inhomogeneities might be similarly insignificant.

Nonetheless, as we have already argued, introducing curvature and inhomogeneities can only reduce the probability of inflation, so even if this does make a difference, the results of Chap. 7 will still stand as an upper bound on the probability of inflation. Thus, relaxing these assumptions cannot change the qualitative conclusion that inflation is very unlikely with this measure.

The door to Plato's academy in Athens was famously engraved with the inscription "Μηδείζ ἀγεωμέτρητοζ εἰσίτω μου τὴν στέγην" (Let no man ignorant of geometry enter). Throughout the centuries, philosophers, mathematicians and physicists have appealed to geometry time and time again for guidance in understanding complex phenomena. As we have seen throughout this thesis the role of geometry in physics is far from over. Even today, over 2,000 years after Plato's academy shut its doors, the modern theories of quantum field theory, particle physics and cosmology still have a place for geometry.

References

1. Vilkovisky GA (1984) The unique effective action in quantum field theory. Nucl Phys B234:125–137 (cit. on pp. 50, 51, 55, 56, 63, 64, 67-69, 79, 90, 91, 94, 96, 124, 243)
2. Vilkovisky GA (1984) The gospel according to DeWitt, pp 169–209 (cit. on pp. 50, 51, 55, 56, 63, 64, 67-69, 79, 90, 91, 94, 96, 124, 243)
3. DeWitt BS (1987) The effective action. In: Architecture of fundamental interactions at short distances: proceedings, Les Houches 44th summer school of theoretical physics: Les Houches, France, July 1–August 8, 1985, pt2, pp 1023–1058 (cit. on pp. 67–69, 105, 124, 243)
4. Falls K, Herrero-Valea M (2019) Frame (In)equivalence in quantum field theory and cosmology. Eur Phys J C 79(7):595. arXiv: 1812.08187 [hep-th] (cit. on pp. 30, 83, 84, 86, 97, 245)
5. Wess J, Zumino B (1974) Supergauge transformations in four-dimensions. Nucl Phys B 70:39–50. Salam A, Sezgin E (eds) (cit. on pp. 107, 113, 247)
6. De Felice A, Tsujikawa S (2010) f(R) theories. Living Rev Rel 13:3. arXiv: 1002.4928 [gr-qc] (cit. on p. 247)
7. Bekenstein JD (1993) The relation between physical and gravitational geometry. Phys Rev D 48:3641–3647. arXiv:gr-qc/9211017 (cit. on p. 247)

8. Bettoni D, Liberati S (2013) Disformal invariance of second order scalar-tensor theories: framing the Horndeski action. Phys Rev D 88:084020. arXiv: 1306.6724 [gr-qc] (cit. on p. 247)
9. Alinea AL, Kubota T (2020) Transformation of primordial cosmological perturbations under the general extended disformal transformation. arXiv: 2005.12747 [gr-qc] (cit. on p. 247)
10. Gildener E (1976) Gauge symmetry hierarchies. Phys Rev D14:1667 (cit. on p. 249)
11. Weinberg S (1979) Gauge hierarchies. Phys Lett 82B:387–391 (cit. on p. 249)
12. Martin J (2012) Everything you always wanted to know about the cosmological constant problem (but were afraid to ask). Comptes Rendus Physique 13:566–665. arXiv: 1205.3365 [astro-ph.CO] (cit. on pp. 162, 233, 249)
13. Sola J (2013) Cosmological constant and vacuum energy: old and new ideas. J Phys Conf Ser 453:012015. Papakostas T, Pliakis DA (eds) arXiv: 1306.1527 [gr-qc] (cit. on pp. 162, 233, 249)
14. Frusciante N, Perenon L (2020) Effective field theory of dark energy: a review. Phys Rept 857:1–63. arXiv: 1907.03150 [astro-ph.CO] (cit. on pp. 162, 233, 249)
15. Carroll SM (2001) The cosmological constant. Living Rev Rel 4:1. arXiv: astro-ph/0004075 (cit. on pp. 162, 233, 249)
16. Weinberg S (1989) The cosmological constant problem. Rev Mod Phys 61:1–23. Hsu J-P, Fine D (eds) (cit. on p. 249)
17. Weinberg S (1987) Anthropic bound on the cosmological constant. Phys Rev Lett 59:2607 (cit. on p. 249)
18. Garriga J, Vilenkin A (2000) On likely values of the cosmological constant. Phys Rev D 61:083502. arXiv: astro-ph/9908115 (cit. on p. 249)
19. Weinberg S (2000) The cosmological constant problems. In: 4th international symposium on sources and detection of dark matter in the universe (DM 2000). arXiv: astro-ph/0005265 (cit. on p. 249)
20. Guth AH (1981) The inflationary universe: a possible solution to the horizon and flatness problems. Phys Rev D23:347–356 (cit. on pp. 154, 184, 249)
21. Linde AD (1982) A new inflationary universe scenario: a possible solution of the horizon, flatness, homogeneity, isotropy and primordial monopole problems. Phys Lett 108B:389–393 (cit. on pp. 154, 184, 249)
22. Albrecht A, Steinhardt PJ (1982) Cosmology for grand unified theories with radiatively induced symmetry breaking. Phys Rev Lett 48:1220–1223 (cit. on pp. 154, 184, 249)
23. Renaux-Petel S, Turzyński K (2016) Geometrical destabilization of inflation. Phys Rev Lett 117(14):141301. arXiv: 1510.01281 [astro-ph.CO] (cit. on p. 250)
24. Renaux-Petel S, Turzyński K, Vennin V (2017) Geometrical destabilization, premature end of inflation and Bayesian model selection. JCAP 11:006. arXiv: 1706.01835 [astro-ph.CO] (cit. on p. 250)
25. Grocholski O, Kalinowski M, Kolanowski M et al (2019) On backreaction effects in geometrical destabilisation of inflation. JCAP 05:008. arXiv: 1901.10468 [astro-ph.CO] (cit. on p. 250)
26. Seery D, Lidsey JE (2005) Primordial non-Gaussianities from multiple-field inflation. JCAP 0509:011. arXiv:astro-ph/0506056 [astro-ph] (cit. on p. 250)
27. Arroja F, Mizuno S, Koyama K (2008) Non-gaussianity from the bispectrum in general multiple field inflation. JCAP 0808:015. arXiv:0806.0619 [astro-ph] (cit. on p. 250)
28. Elliston J, Seery D, Tavakol R (2012) The inflationary bispectrum with curved field-space. JCAP 1211:060. arXiv:1208.6011 [astro-ph.CO] (cit. on p. 250)
29. Garcia-Saenz S, Renaux-Petel S, Ronayne J (2018) Primordial fluctuations and non-Gaussianities in sidetracked inflation. JCAP 07:057. arXiv: 1804.11279 [astro-ph.CO] (cit. on p. 250)
30. Garcia-Saenz S, Renaux-Petel S (2018) Flattened non-Gaussianities from the effective field theory of inflation with imaginary speed of sound. JCAP 11:005. arXiv:1805.12563 [hep-th] (cit. on p. 250)
31. Fumagalli J, Garcia-Saenz S, Pinol L, Renaux-Petel S, Ronayne J (2019) Hyper-non-Gaussianities in inflation with strongly nongeodesic motion. Phys Rev Lett 123(20):201302. arXiv:1902.03221 [hep-th] (cit. on p. 250)

32. Brown AR (2018) Hyperbolic inflation. Phys Rev Lett 121(25):251601. arXiv: 1705.03023 [hep-th] (cit. on p. 250)
33. Mizuno S, Mukohyama S (2017) Primordial perturbations from inflation with a hyperbolic field-space. Phys Rev D 96(10):103533. arXiv:1707.05125 [hep-th] (cit. on p. 250)
34. Bjorkmo T, David Marsh MC (2019) Hyperinflation generalised: from its attractor mechanism to its tension with the 'swampland conditions'. JHEP 04:172. arXiv:1901.08603 [hep-th] (cit. on p. 250)
35. Gibbons GW, Hawking SW, Stewart JM (1987) A natural measure on the set of all universes. Nucl Phys B281:736 (cit. on pp. 197, 198, 250, 251)
36. Gibbons GW, Turok N (2008) The measure problem in cosmology. Phys Rev D77:063516. arXiv: hep-th/0609095 [hep-th] (cit. on pp. 199, 250, 251)
37. Li M, Wang Y (2007) The measure for the multiverse and the probability for inflation. JCAP 0706:012. arXiv: 0704.1026 [hep-th] (cit. on pp. 199, 250)
38. Akrami Y et al (2018) Planck 2018 results. X. Constraints on inflation. arXiv: 1807.06211 [astro-ph.CO] (cit. on pp. 190, 233, 252)
39. Ooguri H, Vafa C (2007) On the geometry of the string landscape and the swampland. Nucl Phys B 766:21–33. arXiv: hep-th/0605264 (cit. on pp. 189, 252)
40. Laplace PS (1952) A philosophical essay on probabilities. Dover books on science. Dover Publications (cit. on pp. 196, 253)

Appendix
Example Covariant Calculations

In this appendix we calculate quantum corrections for a number of example theories. We show explicitly how these calculations can be parametrisation dependent as well as how this dependence can be eliminated with the covariant techniques discussed in the body of the thesis.

A.1 Complex Scalar Field

We start by considering a single complex scalar field ϕ with action

$$S = \int d^4x \left[\partial_\mu \phi \partial^\mu \phi - m^2 |\phi|^2 - \lambda |\phi|^4\right]. \tag{A.1}$$

We have already discussed this theory in Chaps. 1 and 2. However, for completeness, we shall repeat the analysis so that this appendix can be fully understood on its own.

As before, we shall choose $m^2 < 0$ so that the vacuum is

$$\langle \phi \rangle = \frac{1}{\sqrt{2}} \rho_0 \equiv \sqrt{\frac{-m^2}{2\lambda}}. \tag{A.2}$$

Thus, the $U(1)$ symmetry $\phi \to e^{i\theta}\phi$ is spontaneously broken.

A.1.1 Standard Approach: Linear Parametrisation

As we have seen before, we can parametrise the complex field ϕ in terms of its real and imaginary parts as

K. Finn, *Geometric Approaches to Quantum Field Theory*, Springer Theses,
https://doi.org/10.1007/978-3-030-85269-6

$$\phi = \frac{1}{\sqrt{2}}\left(\phi_1 + i\phi_2\right). \tag{A.3}$$

In this parametrisation, the action (A.1) is

$$S = \int d^4x \left[\frac{1}{2}\partial_\mu\phi_1\partial^\mu\phi_1 + \frac{1}{2}\partial_\mu\phi_2\partial^\mu\phi_2 - \frac{1}{2}m^2(\phi_1^2+\phi_2^2) - \frac{\lambda}{4}(\phi_1^2+\phi_2^2)^2\right], \tag{A.4}$$

and the vacuum (A.2) is

$$\langle\phi_1\rangle = \rho_0, \qquad \langle\phi_2\rangle = 0. \tag{A.5}$$

We will calculate perturbations around this vacuum, setting

$$\phi_1 = \rho_0 + \delta\phi_1, \qquad \phi_2 = 0 + \delta\phi_2. \tag{A.6}$$

In accordance with Goldstone's theorem [1–3], the perturbations consist of a massive Higgs mode, corresponding to $\delta\phi_1$, and a massless Goldstone mode corresponding to $\delta\phi_2$.

We have seen in Sect. 2.4.1 that quantum corrections to this theory can be calculated using Feynman diagrams. We can derive the Feynman rules for (A.4) using (2.79) and (2.80) and we find

 $= \dfrac{i}{p^2 - m_1^2}$, $= \dfrac{i}{p^2}$,

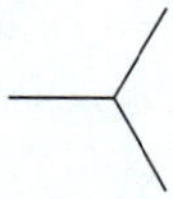 $= -6i\lambda\rho_0$,  $= -2i\lambda\rho_0$, (A.7)

 $= -6i\lambda$,  $= -6i\lambda$,  $= -2i\lambda$,

where

$$m_1^2 \equiv m^2 + 3\lambda\rho_0^2 = -2m^2 \tag{A.8}$$

is the mass of the Higgs mode. Here, we represent $\delta\phi_1$ by a solid line and $\delta\phi_2$ by a dashed line.

Let us use these Feynman rules to calculate the renormalisation of the Higgs mass. At one loop order we have

$i\Gamma_{\phi_1\phi_1}(p) =$ $+$ 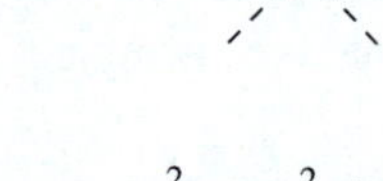 $+$  $+$

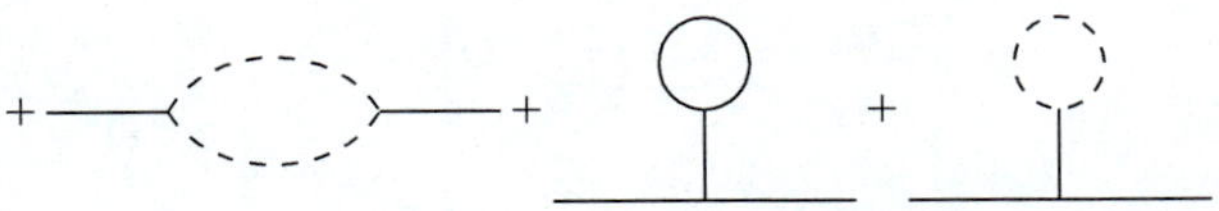

$$
\begin{aligned}
&= i(p^2 - m_1^2) + \frac{3i\lambda}{(4\pi)^2} A(m_1^2) + \frac{i\lambda}{(4\pi)^2} A(0) + 18i \frac{\lambda^2 \rho_0^2}{(4\pi)^2} B_0(p^2, m_1, m_1) \\
&+ 2i \frac{\lambda^2 \rho_0^2}{(4\pi)^2} B_0(p^2, 0, 0) - 18i \frac{\lambda^2 \rho_0^2}{(4\pi)^2 m_1^2} A(m_1^2) - 6i \frac{\lambda^2 \rho_0^2}{(4\pi)^2 m_1^2} A(0).
\end{aligned} \tag{A.9}
$$

Here, we have defined the following two integrals

$$
A(m^2) \equiv \int \frac{d^4k}{i\pi^2} \frac{1}{k^2 - m^2}, \tag{A.10}
$$

$$
B_0(p^2, m_1, m_2) \equiv \int \frac{d^4k}{i\pi^2} \frac{1}{k^2 - m_1^2} \frac{1}{(p+k)^2 - m_2^2}, \tag{A.11}
$$

which we can perform to give

$$
A(m^2) = m^2 \left[C_{\mathrm{UV}} + 1 - \ln\left(\frac{m^2}{\mu^2} \right) \right], \tag{A.12}
$$

$$
B_0(p^2, m_1, m_2) = C_{\mathrm{UV}} - \int_0^1 dx \ln\left(\frac{m_1^2(1-x) + m_2^2 x - x(1-x)p^2}{\mu^2} \right). \tag{A.13}
$$

Here, μ is the renormalisation scale,

$$
C_{\mathrm{UV}} = \frac{2}{4 - D} - \gamma_{\mathrm{E}} + \ln(4\pi) \tag{A.14}
$$

is the UV divergence that is cancelled by counterterms in the $\overline{\mathrm{MS}}$ renormalisation scheme [4, 5], and $\gamma_{\mathrm{E}} = 0.577\ldots$ is the Euler–Mascheroni constant. We have performed the calculation in $D = 4 - 2\epsilon$ dimensions using dimensional regularisation [6].

We, therefore, have

$$
\begin{aligned}
\Gamma_{\phi_1\phi_1}(p) =& (p^2 - m_1^2) + \frac{\lambda m_1^2}{4\pi^2} \ln\left(\frac{p^2}{\mu^2} \right) \\
&- \frac{\lambda m_1^2}{(4\pi)^2} \left[-4C_{\mathrm{UV}} + 4 + 9 \int_0^1 dx \ln\left(\frac{x(x-1)p^2 + m_1^2}{\mu^2} \right) \right].
\end{aligned} \tag{A.15}
$$

Note that $A(0) = 0$ and, thus, the third and final diagrams in (A.9) give no contribution. This is the same calculation that was performed in Chap. 1.

From (A.15) we see that there is no wavefunction renormalisation and the beta function for the Higgs mass is

$$\beta_{m_1^2} = -\mu \frac{\partial \widehat{\Gamma}_2}{\partial \mu} = \frac{\lambda m_1^2}{2\pi^2}. \tag{A.16}$$

We can also calculate the one-loop Goldstone self energy

$i\Gamma_{\phi_2\phi_2}(p) =$

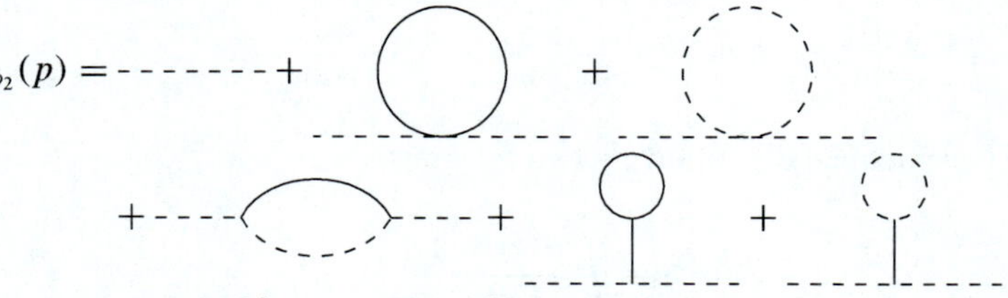

$$\begin{aligned} &= ip^2 + \frac{i\lambda}{(4\pi)^2} A(m_1^2) + \frac{3i}{(4\pi)^2} \lambda A(0) \\ &+ 4i \frac{\lambda^2 \rho_0^2}{(4\pi)^2} B_0(p^2, m_1, 0) - 6i \frac{\lambda^2 \rho_0^2}{(4\pi)^2 m_1^2} A(m_1^2) - 2i \frac{\lambda^2 \rho_0^2}{(4\pi)^2 m_1^2} A(0) \\ &= ip^2 - \frac{2i\lambda m_1^2}{16\pi^2} \int_0^1 dx \ln\left(1 - \frac{xp^2}{m_1^2}\right). \end{aligned} \tag{A.17}$$

Since (A.17) has no dependence on μ, the Goldstone mass is not renormalised and remains zero in accordance with Goldstone's theorem.

We can also calculate quantum effects for this theory using the quantum effective action as discussed in Sect. 2.4.2. The one-loop effective action can be calculated from the inverse propagator, which for the theory (A.4) is

$$\begin{aligned} &\frac{\delta^2 S}{\delta\phi^A(x)\delta\phi^B(y)} \\ &= \begin{pmatrix} -\partial^2 - m^2 - 3\lambda\phi_1^2 - \lambda\phi_2^2 & -2\lambda\phi_1\phi_2 \\ -2\lambda\phi_1\phi_2 & -\partial^2 - m^2 - 3\lambda\phi_2^2 - \lambda\phi_1^2 \end{pmatrix} \delta^{(4)}(x-y). \end{aligned} \tag{A.18}$$

Without loss of generality, we can use the $U(1)$ symmetry to set $\phi_2 = 0$. In addition, we shall focus on a static configuration of the fields, which will allow us to calculate the effective potential from (2.92). We find

$$\begin{aligned} V_{\text{eff}}(\varphi) \equiv &- \frac{1}{V_4} \Gamma[\phi_1 = \varphi, \phi_2 = 0] \\ =&V(\varphi) + \frac{i}{2} \ln\det\left[\partial^2 + m^2 + 3\lambda\varphi^2\right] + \frac{i}{2} \ln\det\left[\partial^2 + m^2 + \lambda\varphi^2\right] \\ =&\frac{1}{2} m^2\varphi^2 + \frac{1}{4}\lambda\varphi^4 + \frac{1}{64\pi^2}\left\{(m^2 + 3\lambda\varphi^2)^2\left[\ln\left(\frac{m^2 + 3\lambda\varphi^2}{\mu^2}\right) - \frac{3}{2}\right]\right. \\ &\left. + (m^2 + \lambda\varphi^2)^2\left[\ln\left(\frac{m^2 + \lambda\varphi^2}{\mu^2}\right) - \frac{3}{2}\right]\right\}. \end{aligned} \tag{A.19}$$

Here, V_4 is the total four-volume of spacetime and we have again used the $\overline{\text{MS}}$ renormalisation scheme. This result was previously shown in (1.16).

Finally, let us compute the renormalisation of the coupling constant λ. The simplest way to perform this calculation is using the Callan Symanzic equation [7, 8], which tells us

$$\left[\mu\frac{\partial}{\partial\mu} + \beta_\lambda\frac{\partial}{\partial\lambda} + \beta_{m^2}\frac{\partial}{\partial m^2}\right]\widetilde{V}_{\text{eff}} = 0. \tag{A.20}$$

where

$$\widetilde{V}_{\text{eff}}(\varphi) = V_{\text{eff}}(\varphi) - V_{\text{eff}}(0) \tag{A.21}$$

is the modified effective potential.

Plugging in the expression for V_{eff} from (A.19) we find, at leading order,

$$-\frac{(m^2+3\lambda\varphi^2)^2 + (m^2+\lambda\varphi^2)^2 - 2m^4}{32\pi^2} + \frac{1}{4}\beta_\lambda\varphi^4 - \frac{1}{4}\beta_{m_1^2}\varphi^2 = 0. \tag{A.22}$$

Here, we have used the identity $\beta_{m_1^2} = -2\beta_{m^2}$, which derives from (A.8).

We can rearrange (A.22) and plug in the expression for $\beta_{m_1^2}$ from (A.16). We, therefore, calculate the beta function of the coupling constant, evaluated at the vacuum $\varphi = \rho_0$, to be

$$\beta_\lambda = \frac{5}{4\pi^2}\lambda^2. \tag{A.23}$$

A.1.2 Standard Approach: Non-linear Parametrisation

Another way of parametrising the complex field is through its modulus and argument:

$$\phi = \frac{1}{\sqrt{2}}\rho e^{i\frac{\sigma}{\rho_0}}. \tag{A.24}$$

Expressed in terms of this parametrisation, the action (A.1) becomes

$$S = \int d^4x\left[\frac{1}{2}\partial_\mu\rho\partial^\mu\rho + \frac{1}{2}\left(\frac{\rho}{\rho_0}\right)^2\partial_\mu\sigma\partial^\mu\sigma - \frac{1}{2}m^2\rho^2 - \frac{\lambda}{4}\rho^4\right] \tag{A.25}$$

and the vacuum (A.2) becomes

$$\langle\rho\rangle = \rho_0, \qquad \langle\sigma\rangle = 0. \tag{A.26}$$

We again take perturbations around this vacuum and, thus, define

$$\rho = \rho_0 + \delta\rho, \qquad\qquad \sigma = 0 + \delta\sigma. \tag{A.27}$$

As discussed in Chap. 1, we have

$$\delta\rho = \delta\phi_1 + O(\delta\phi^2), \qquad \delta\sigma = \delta\phi_2 + O(\delta\phi^2) \tag{A.28}$$

and, thus, these correspond to the same perturbations as we considered previously.

As before, we can use (2.79) and (2.80) to calculate the Feynman rules for this theory and we find

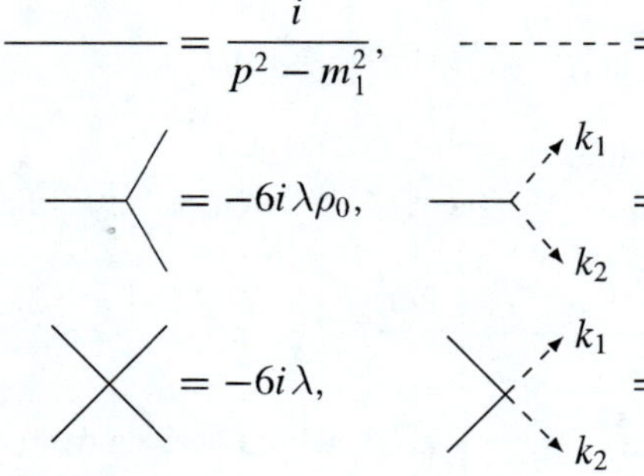

$$= \frac{i}{p^2 - m_1^2}, \qquad = \frac{i}{p^2},$$
$$= -6i\lambda\rho_0, \qquad = -\frac{2i}{\rho_0} k_1 \cdot k_2, \tag{A.29}$$
$$= -6i\lambda, \qquad = -\frac{2i}{\rho_0^2} k_1 \cdot k_2.$$

Here, the solid line corresponds to $\delta\rho$ and the dashed line to $\delta\sigma$. We see that (A.29) differs from (A.7), showing explicitly the parametrisation non-invariance of the standard approach to Feynman rules.

We can calculate the Higgs mass renormalisation using this parametrisation, and we find

$i\Gamma_{\rho\rho}(p) =$

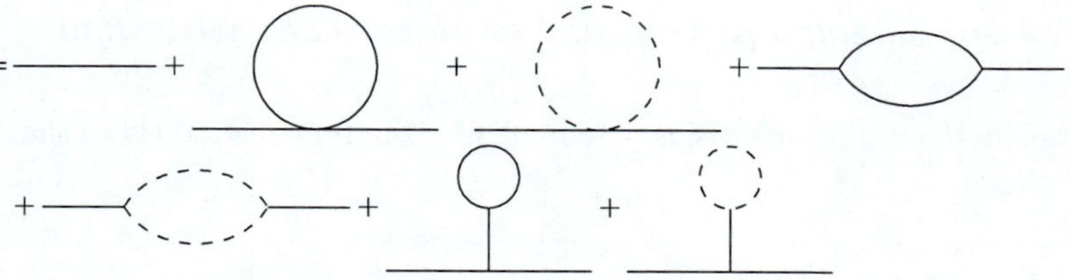

$$\begin{aligned}
&= i(p^2 - m_1^2) + \frac{3i\lambda}{(4\pi)^2} A(m_1^2) + \frac{2i}{\rho_0} \int \frac{d^4k}{(2\pi)^4} + 18i \frac{\lambda^2 \rho_0^2}{(4\pi)^2} B_0(p^2, m_1, m_1) \\
&\quad + \frac{i}{2} \frac{p^4}{(4\pi)^2 \rho_0^2} B_0(p^2, 0, 0) - 18i \frac{\lambda^2 \rho_0^2}{(4\pi)^2 m_1^2} A(m_1^2) + 6i \frac{\lambda^2 \rho_0^2}{m_1^2} \int \frac{d^4k}{(2\pi)^4} \\
&= i(p^2 - m_1^2) + \frac{3\lambda m_1^2 + \lambda \frac{p^4}{m_1^2}}{(4\pi)^2} \ln\left(\frac{p^2}{\mu^2}\right) \\
&\quad + \frac{i\lambda m_1^2}{(4\pi)^2} \left[\left(3 + \frac{p^4}{m_1^4}\right) C_{\mathrm{UV}} - 6 + 2\frac{p^4}{m_1^4} - 9 \int_0^1 dx \ln\left(\frac{x(x-1)p^2 + m_1^2}{\mu^2}\right) \right].
\end{aligned} \tag{A.30}$$

This calculation was previously shown in (1.23). We see that the Higgs mass renormalisation calculated in terms of ρ and σ differs from the expression calculated using ϕ_1 and ϕ_2 despite the fact that the underlying theory is the same in both cases.

In fact, due to the presence of the p^4 divergence, (A.30) is naively non-renormalisable while (A.16) is straightforwardly renormalisable.

We note, however, that when we consider on shell momentum, so that $p^2 = m_1^2$, the two expressions, (A.15) and (A.30) are equal. The on shell beta function for the Higgs mass will, therefore, be identical in both cases.

Next we calculate the renormalisation of the Goldstone mass

$i\Gamma_{\sigma\sigma}(p) =$

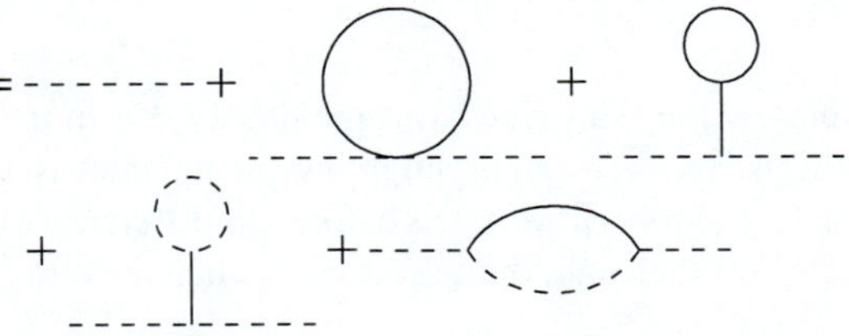

$$
\begin{aligned}
&= ip^2 - \frac{ip^2}{(4\pi)^2\rho_0^2} A(m_1^2) + 6i\frac{\lambda p^2}{(4\pi)^2 m_1^2} A(m_1^2) \\
&\quad - 2i\frac{p^2}{\rho_0^2}\int \frac{d^4k}{(2\pi)^4} + \left[i\frac{3p^2 - m_1^2}{(4\pi)^2\rho_0^2} A(m_1^2) + i\frac{(p^2 - m_1^2)^2}{(4\pi)^2\rho_0^2} B_0(p^2, m_1, 0) \right] \\
&= ip^2 + i\frac{5p^2 - m_1^2}{16\pi^2\rho_0^2} m_1^2 \left[C_{\rm UV} + 1 - \ln\left(\frac{m^2}{\mu^2}\right) \right] \\
&\quad + i\frac{(p^2 - m_1^2)^2}{16\pi^2\rho_0^2} \left[C_{\rm UV} - \int_0^1 dx \ln\left(\frac{(1-x)m_1^2 - x(1-x)p^2}{\mu^2} \right) \right].
\end{aligned} \tag{A.31}
$$

This differs from (A.17) and so we again get a different result using ρ and σ than we did using ϕ_1 and ϕ_2. We also see that, just like the Higgs mass, (A.31) contains terms that are naively non-renormalisable.

However, as before, this difference is only present off shell. On shell, when $p^2 = 0$, we have

$$
\Gamma_{\sigma\sigma}(p^2 = 0) = 0, \tag{A.32}
$$

in agreement with (A.17). Therefore, the Goldstone mass is not renormalised as expected by Goldstone's theorem.

Let us also calculate the quantum effective action in this parametrisation. We start by calculating the inverse propagator, which gives

$$
\begin{aligned}
&\frac{\delta^2 S}{\delta\phi^a(x)\delta\phi^b(y)} \\
&= \begin{pmatrix} -\partial^2 + \partial_\mu\sigma\partial^\mu\sigma - m^2 - 3\lambda\rho^2 & -2\partial_\mu\rho\partial^\mu\sigma - 2\rho\partial^2\sigma - 2\rho\partial_\mu\sigma\partial^\mu \\ -2\partial_\mu\rho\partial^\mu\sigma - 2\rho\partial^2\sigma - 2\rho\partial_\mu\sigma\partial^\mu & -2\rho\partial_\mu\rho\partial^\mu - \rho^2\partial^2 \end{pmatrix} \delta^{(4)}(x-y).
\end{aligned} \tag{A.33}
$$

We can then calculate the effective potential by considering a static configuration. As before we can use the $U(1)$ symmetry to set $\sigma = 0$ without loss of generality. We, therefore, find

$$\begin{aligned}V_{\rm eff}(\varphi) &\equiv -\frac{1}{V_4}\Gamma[\rho=\varphi,\sigma=0]\\&=\frac{1}{2}m^2\varphi^2+\frac{\lambda}{4}\varphi^4-\ln\det\left[\partial^2+m^2+3\lambda\varphi^2\right]-\ln\det\left[\varphi^2\partial^2\right]\\&=\frac{1}{2}m^2\varphi^2+\frac{\lambda}{4}\varphi^4+\frac{1}{64\pi^2}\left\{(m^2+3\lambda\varphi^2)^2\left[\ln\left(\frac{m^2+3\lambda\varphi^2}{\mu^2}\right)-\frac{3}{2}\right]\right\}\end{aligned} \tag{A.34}$$

in the $\overline{\rm MS}$ renormalisation scheme. This result was previously shown in (1.24).

We see that (A.19) differs from (A.34), highlighting the parametrisation dependence of the standard effective action. However, as before, the difference only appears off shell. If we put the fields on shell and, thus, set $\varphi=\rho_0$, then the two expressions agree.

Before proceeding to calculate the coupling renormalisation we notice the following. We have so far been considering the case $m^2<0$ with the symmetry broken vacuum (A.2). However, we can also consider taking $m^2>0$, for which the vacuum lies at $\varphi=0$. Surprisingly, in this case (A.19) and (A.34) do not agree even on shell.

The reason for this unexpected behaviour is that the coordinate chart (A.24) does not cover the point $\phi=0$. As discussed in Chap. 2, to cover the entire field space manifold we sometimes need an atlas consisting of multiple charts. To consider the point $\phi=0$ we must, therefore, switch to a different chart. For example, we can use an offset parametrisation $\phi=\frac{1}{\sqrt{2}}\left(\tilde{\rho}e^{i\tilde{\sigma}}-\delta\right)$, where δ is a constant. This chart now covers the vacuum $\phi=0$.

In the offset parametrisation, the effective potential is

$$\begin{aligned}V_{\rm eff}(\tilde{\rho}=\tilde{\varphi},\tilde{\sigma}=0)=&\frac{1}{2}m^2(\tilde{\varphi}-\delta)^2+\frac{\lambda}{4}(\tilde{\varphi}-\delta)^4\\&+\frac{(m^2+3\lambda(\tilde{\varphi}-\delta)^2)^2}{64\pi^2}\left[\ln\left(\frac{m^2+3\lambda(\tilde{\varphi}-\delta)^2}{\mu^2}\right)-\frac{3}{2}\right]\\&+\frac{1}{64\pi^2}\frac{\delta^2}{\tilde{\varphi}^2}(m^2+\lambda(\tilde{\varphi}-\delta)^2)^2\left[\ln\left(\frac{\delta}{\tilde{\varphi}}\frac{m^2+\lambda(\tilde{\varphi}-\delta)^2}{\mu^2}\right)-\frac{3}{2}\right].\end{aligned} \tag{A.35}$$

We see that when evaluated at the symmetry unbroken vacuum, which now corresponds to $\tilde{\varphi}=\delta$, the two expressions (A.35) and (A.19) agree.

Finally, we consider the renormalisation of the coupling constant using the Callan–Symanzic equation as before. In order to compute the modified potential we need to subtract from (A.34) the effective potential evaluated at $\phi=0$. As we have seen, this point is not covered by the coordinate chart (A.24) and so the correct way to calculate this offset is to use (A.35) and take $\tilde{\varphi}=\delta$. This gives us

$$V_{\rm eff}(0)=\frac{2m^2}{64\pi^2}\left[\ln\left(\frac{m^2}{\mu^2}\right)-\frac{3}{2}\right]. \tag{A.36}$$

The Callan–Symanzic equation in this parametrisation, therefore, reduces to

$$-\frac{(m^2+3\lambda\varphi^2)^2-2m^4}{32\pi^2}+\frac{1}{4}\beta_\lambda\varphi^4-\frac{1}{4}\beta_{m_1^2}\varphi^2=0. \quad \text{(A.37)}$$

On shell we have $\varphi=\rho_0$ and this becomes identical to (A.22), as expected. Thus, the on shell beta function for the coupling renormalisation is (A.23) as before.

We see that calculations using the two different parametrisations led to several differences in the intermediate, off shell results. However, the above calculations gave the same result for all physical, on shell observables, regardless of the parametrisation.

A.1.3 Covariant Approach

We have seen in Sect. 2.5 how to alleviate the parametrisation dependence of these calculations and recover manifest reparametrisation invariance. Let us, therefore, apply the techniques of field space covariance to the theory (A.1).

The fields ϕ_1 and ϕ_2 are canonically normalised as seen in (A.4). Thus, the field space is trivial in this parametrisation. There will, therefore, be no difference between the covariant and standard approaches in this case. The VDW effective potential will be (A.19), the covariant Feynman rules will give (A.7), and the renormalisation group calculations will be identical to those in Sect. A.1.1.

We will, therefore, focus on the non-linear parametrisation (A.24). We can calculate the configuration space metric from (A.25) using (2.61) and we find

$$\mathcal{G}_{\widehat{A}\widehat{B}}=\begin{pmatrix}1 & 0\\ 0 & (\rho/\rho_0)^2\end{pmatrix}\delta^{(4)}(x_A-x_B) \quad \text{(A.38)}$$

The configuration-space Christoffel symbols can be calculated from (2.65) and we find

$$\Gamma^{\rho(z)}_{\sigma(x)\sigma(y)}=-\frac{\rho(z)}{\rho_0^2}\delta^{(4)}(z-x)\delta^{(4)}(z-y), \quad \text{(A.39)}$$

$$\Gamma^{\sigma(z)}_{\rho(x)\sigma(y)}=\frac{1}{\rho(z)}\delta^{(4)}(z-x)\delta^{(4)}(z-y), \quad \text{(A.40)}$$

with all others equal to zero.

As shown in Sect. 2.5.1, we can calculate a covariant set of Feynman rules for this theory using the covariant derivative on the configuration space. Performing these derivatives we find

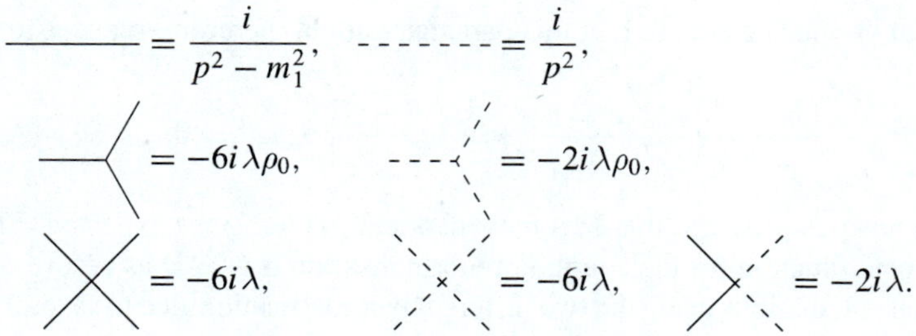

$$= \frac{i}{p^2 - m_1^2}, \quad = \frac{i}{p^2},$$
$$= -6i\lambda\rho_0, \quad = -2i\lambda\rho_0, \tag{A.41}$$
$$= -6i\lambda, \quad = -6i\lambda, \quad = -2i\lambda.$$

These are identical to (A.7) and, thus, any Feynman diagram will give the same results as in Sect. A.1.1 both on and off shell. As expected, the covariant Feynman rules are manifestly reparametrisation invariant.

We can now calculate the VDW effective action. For this we will need the covariant inverse propagator, which we can calculate to be

$$\begin{aligned} \nabla_{\widehat{A}}\nabla_{\widehat{B}}S &= \delta^{(4)}(x_A - x_B) \\ &\times \begin{pmatrix} -\partial^2 + \partial_\mu\sigma\partial^\mu\sigma - m^2 - 3\lambda\rho^2 & \rho\partial_\mu\sigma\partial^\mu \\ \rho\partial_\mu\sigma\partial^\mu & -\rho\partial_\mu\rho\partial^\mu - \rho^2\partial^2 + \rho^2\partial_\mu\sigma\partial^\mu\sigma - m^2\rho^2 - \lambda\rho^4 \end{pmatrix}. \end{aligned} \tag{A.42}$$

As before, we can calculate the effective potential by considering a static configuration and can use the $U(1)$ symmetry to set $\sigma = 0$ without loss of generality. For this configuration we have

$$\nabla_{\widehat{A}}\nabla_{\widehat{B}}S\Big|_{\rho=\varphi,\sigma=0} = \begin{pmatrix} -\partial^2 - m^2 - 3\lambda\varphi^2 & 0 \\ 0 & -\varphi^2\partial^2 - m^2\varphi^2 - \lambda\varphi^4 \end{pmatrix} \delta^{(4)}(x_A - x_B). \tag{A.43}$$

The one-loop VDW effective potential is therefore

$$\begin{aligned} V_{\text{eff}}(\varphi) &\equiv -\frac{1}{V_4}\Gamma[\rho = \varphi, \sigma = 0] \\ &= \frac{1}{2}m^2\varphi^2 + \frac{\lambda}{4}\varphi^4 - \ln\det\left[\partial^2 + m^2 + 3\lambda\varphi^2\right] - \ln\det\left[\partial^2 + m^2 + \lambda\varphi^2\right] \\ &\quad - \ln(\varphi^2) + \ln\det[\mathcal{G}_{\widehat{A}\widehat{B}}] \\ &= \frac{1}{2}m^2\varphi^2 + \frac{1}{4}\lambda\varphi^4 + \frac{1}{64\pi^2}\left\{(m^2 + 3\lambda\varphi^2)^2\left[\ln\left(\frac{m^2 + 3\lambda\varphi^2}{\mu^2}\right) - \frac{3}{2}\right]\right. \\ &\quad \left. + (m^2 + \lambda\varphi^2)^2\left[\ln\left(\frac{m^2 + \lambda\varphi^2}{\mu^2}\right) - \frac{3}{2}\right]\right\}, \end{aligned} \tag{A.44}$$

in the $\overline{\text{MS}}$ scheme. Here, V_4 is the four-volume of spacetime. Note that the term $\ln\det[\mathcal{G}_{\widehat{A}\widehat{B}}]$ is proportional to $\delta^{(4)}(0)$ and is, thus, removed by the regularisation scheme.

The expression (A.44) is identical to (A.19). Thus, the VDW effective action is reparametrisation invariant.

A.2 Curved Field-Space Example

The previous example was special in that there existed a parametrisation of the fields in which the kinetic terms were canonical. Thus, the field space manifold in that case was flat. We now wish to consider an example with a curved field space to investigate the effects of field space curvature on quantum calculations. We will consider a simple theory with two fields ρ and σ, and will take σ to be an angular variable with a shift symmetry as before. In addition, we require that the metric of the field space be positive-definite in order to avoid tachyonic instabilities [9]. Thus, we consider a field space metric of the form

$$G_{AB} = \begin{pmatrix} 1 & 0 \\ 0 & \left(\frac{\rho}{\rho_0}\right)^{2n} \end{pmatrix}. \tag{A.45}$$

We can calculate the Christoffel symbols of this metric and find them to be

$$\Gamma^{\rho}_{\sigma\sigma} = -n\frac{\rho^{2n-1}}{\rho_0^{2n}}, \qquad \Gamma^{\sigma}_{\rho\sigma} = \frac{n}{\rho}, \tag{A.46}$$

with all others equal to zero. We can then calculate the non-zero components of the field-space Riemann tensor, which are

$$\begin{aligned} R^{\rho}_{\ \sigma\rho\sigma} &= -\frac{n(n-1)}{\rho^2}\frac{\rho^{2n}}{\rho_0^{2n}}, & R^{\rho}_{\ \sigma\sigma\rho} &= \frac{n(n-1)}{\rho^2}\frac{\rho^{2n}}{\rho_0^{2n}}, \\ R^{\sigma}_{\ \rho\rho\sigma} &= \frac{n(n-1)}{\rho^2}, & R^{\sigma}_{\ \rho\sigma\rho} &= -\frac{n(n-1)}{\rho^2}. \end{aligned} \tag{A.47}$$

These vanish for the cases $n = 0$ and $n = 1$, which we have already analysed in Sect. A.1, but are non-zero in all other cases. Thus, this field space is non-trivially curved for all $n \geq 2$.

The simplest model with curvature is, therefore, the case $n = 2$, which has the Lagrangian

$$\mathcal{L} = \frac{1}{2}\partial_\mu\rho\partial^\mu\rho + \frac{1}{2}\left(\frac{\rho}{\rho_0}\right)^4 \partial_\mu\sigma\partial^\mu\sigma - \frac{1}{2}m^2\rho^2 - \frac{\lambda}{4}\rho^4. \tag{A.48}$$

As before, we shall consider the symmetry broken vacuum

$$\langle\rho\rangle = \rho_0 \equiv \sqrt{\frac{-m^2}{\lambda}}. \tag{A.49}$$

A.2.1 Standard Approach

We start by calculating the standard Feynman rules for the Lagrangian (A.48). These are given by

$$= \frac{i}{p^2 - m_1^2},$$

$$= \frac{i}{p^2},$$

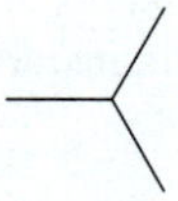

$$= -6i\lambda\rho_0,$$

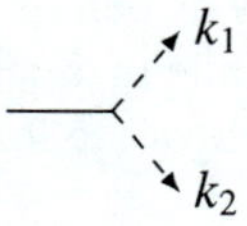

$$= -4i\frac{k_1 \cdot k_2}{\rho_0}, \tag{A.50}$$

$$= -6i\lambda,$$

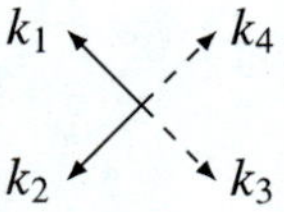

$$= -12i\frac{k_3 \cdot k_4}{\rho_0^2},$$

where $m_1^2 \equiv m^2 + 3\lambda\rho_0^2$ as before. There are also higher order interactions, which we have not calculated since they will not be needed for one loop calculations.

We can now calculate the self-energy of the ρ field. This is given by

$i\Gamma_{\rho\rho}(p) =$ $+$ 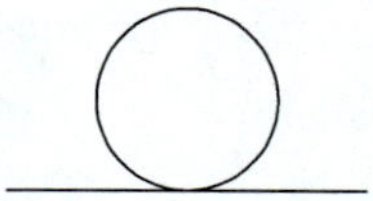 $+$  $+$ 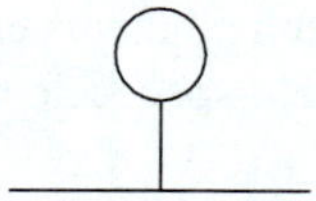

$+$ 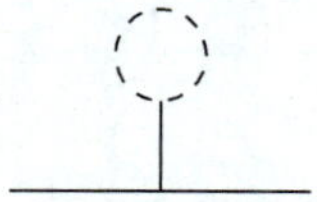 $+$  $+$ 

$$\begin{aligned}
&= i(p^2 - m_1^2) + \frac{3i\lambda}{(4\pi)^2}A(m_1^2) + 18i\frac{\lambda^2\rho_0^2}{(4\pi)^2}B_0(p^2, m_1, m_1) - 18i\frac{\lambda^2\rho_0^2}{(4\pi)^2 m_1^2}A(m_1^2) \\
&\quad + 12\frac{\lambda}{m_1^2}\int\frac{d^4k}{(2\pi)^4} + -6\frac{1}{\rho_0^2}\int\frac{d^4k}{(2\pi)^4} + \frac{2ip^4}{(4\pi)^2\rho_0^2}B_0(p^2, 0, 0) \\
&= i(p^2 - m_1^2) + \frac{i\lambda m_1^2}{(4\pi)^2}\left[\left(3 + 4\frac{p^4}{m_1^4}\right)C_{\text{UV}} + 6\ln\left(\frac{m_1^2}{\mu^2}\right) - 4\frac{p^4}{m_1^4}\ln\left(\frac{p^2}{\mu^2}\right)\right. \\
&\qquad \left. -9\int_0^1 dx\ln\left(\frac{m_1^2 - x(1-x)p^2}{\mu^2}\right) - 6 + 8\frac{p^4}{m_1^4}\right].
\end{aligned} \tag{A.51}$$

We see there is a UV-divergent term proportional to p^4. This term cannot be absorbed into a counterterm and, thus, the theory (A.48) is non-renormalisable. Nevertheless we shall proceed regardless and define the on shell self energy

$$\Gamma_{\rho\rho}(p^2 = m_1^2) = \frac{\lambda m_1^2}{(4\pi)^2}\left[-7C_{\mathrm{UV}} - 7\ln\left(\frac{m_1^2}{\mu^2}\right) + 20 - 3\sqrt{3}\pi\right]. \tag{A.52}$$

We can also calculate the Goldstone self-energy. This is given by

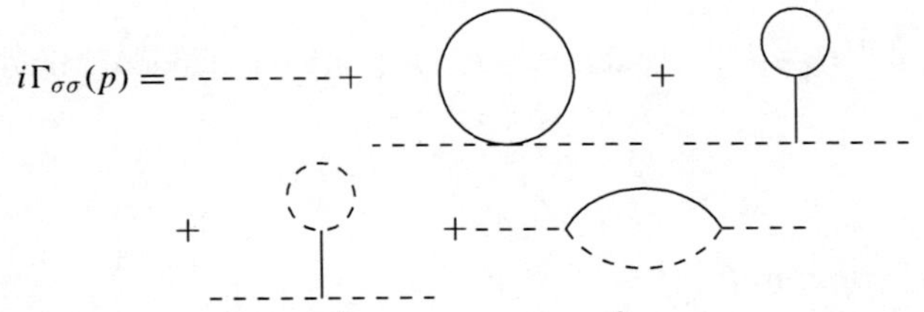

$$\begin{aligned} &= ip^2 + \frac{6ip^2}{(4\pi)^2\rho_0^2}A(m_1^2) - 12i\frac{p^2\lambda}{(4\pi)^2 m_1^2}A(m_1^2) \\ &\quad - 16\frac{p^2}{\rho_0^2 m_1^2}\int\frac{d^4k}{(2\pi)^4} + 4i\left\{\frac{3p^2 - m_1^2}{(4\pi)^2\rho_0^2}A(m_1^2) + \frac{(p^2 - m_1^2)^2}{(4\pi)^2\rho_0^2}B_0(p^4, m_1, 0)\right\} \\ &= ip^2 + 4i\frac{3p^2 - m_1^2}{(4\pi)^2\rho_0^2}m_1^2\left[C_{\mathrm{UV}} + 1 - \ln\left(\frac{m_1^2}{\mu^2}\right)\right] \\ &\quad + 4i\frac{(p^2 - m_1^2)^2}{(4\pi)^2\rho_0^2}\left[C_{\mathrm{UV}} + 1 - \int_0^1 dx\ln\left(\frac{xp^2 - m_1^2}{\mu^2}\right)\right]. \end{aligned} \tag{A.53}$$

As before, this expression contains divergences that cannot be absorbed by a counterterm. However, we can still calculate the on shell self energy by setting $p^2 = 0$ and we find

$$\Gamma_{\sigma\sigma}(p^2 = 0) = 0. \tag{A.54}$$

Thus, the Goldstone boson receives no correction to its mass as expected.

Finally, it is instructive to calculate the tree-level S-matrix element for $\rho\rho \to \sigma\sigma$. This can be calculated from the Feynman diagrams

$i\mathcal{M}(\rho\rho \to \sigma\sigma) =$

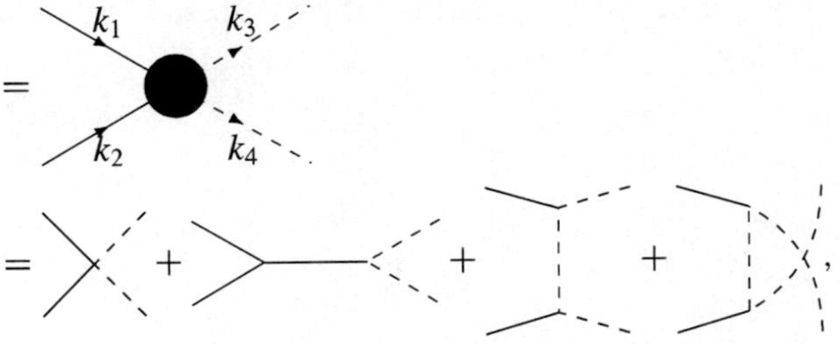

(A.55)

which can be evaluated to give

$$
\begin{aligned}
i\mathcal{M}(\rho\rho \to \sigma\sigma) &= -6i\frac{s}{\rho_0^2} - 6i\frac{sm_1^2}{\rho_0^2(s-m_1^2)} - 4i\frac{(m_1^2-t)^2}{\rho_0^2 t} - 4i\frac{(m_1^2-u)^2}{\rho_0^2 u} \\
&= -\frac{2i}{\rho_0^2}\left[3\frac{s^2}{s-m_1^2} + 2\frac{(m_1^2-t)^2}{t} + 2\frac{(m_1^2-u)^2}{u}\right],
\end{aligned}
\tag{A.56}
$$

where $s = (k_1 + k_2)^2$, $t = (k_1 - k_3)^2$, and $u = (k_1 - k_4)^2$ are the standard Mandelstam variables.

A.2.2 Covariant Approach

Let us now analyse the same theory using the covariant approach. The covariant Feynman rules can be calculated from (2.105) and are found to be

 $= \dfrac{i}{p^2 - m_1^2}$, $= \dfrac{i}{p^2}$,

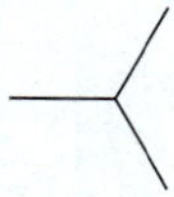 $= -6i\lambda\rho_0$, 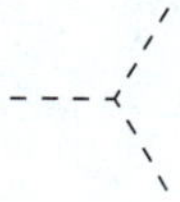 $= -4i\lambda\rho_0$, (A.57)

 $= -6i\lambda$,  $= -24i\lambda$.

There is also an infinite set of higher order vertices, which are not shown here. These do not enter into one-loop calculations and so it is not necessary to derive them.

We have also omitted the $\rho\rho\sigma\sigma$ vertex from (A.57). We wish to highlight this vertex separately because there are different ways we could define this vertex depending on the order in which we take the covariant derivative. Because the field space for this theory has a non-trivial curvature, the covariant derivative does not commute. Thus, the choice of ordering leads to different possible definitions for this vertex. Explicitly, we have

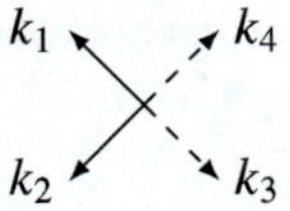

$\rho\rho\sigma\sigma$

$$\equiv \nabla_\rho \nabla_\rho \nabla_\sigma \nabla_\sigma S = -4i\lambda - 4i\frac{k_1 \cdot k_2}{\rho_0^2},$$

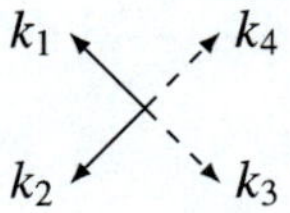

$\sigma\sigma\rho\rho$

$$\equiv \nabla_\sigma \nabla_\sigma \nabla_\rho \nabla_\rho S = 4i\lambda - 4i\frac{k_3 \cdot k_4}{\rho_0^2},$$

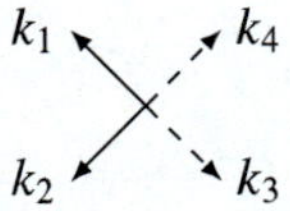

$\rho\sigma\rho\sigma$
$\rho\sigma\sigma\rho$

$$\equiv \nabla_\rho \nabla_\sigma \nabla_\rho \nabla_\sigma S = 2i\frac{k_1 \cdot (k_3 + k_4 - k_2)}{\rho_0^2}, \tag{A.58}$$

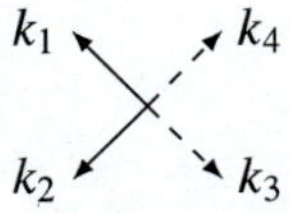

$\sigma\rho\rho\sigma$
$\sigma\rho\sigma\rho$

$$\equiv \nabla_\sigma \nabla_\rho \nabla_\rho \nabla_\sigma S = 4i\lambda + 2i\frac{k_3 \cdot (k_1 + k_2 - k_4)}{\rho_0^2}.$$

The indices after the diagram here denote the ordering. Note that we can also change the order of the two ρ particles or the two σ particles, individually. This is equivalent to exchanging $k_1 \leftrightarrow k_2$ and/or $k_3 \leftrightarrow k_4$.

It is informative to calculate the on shell form of these vertices by setting

$$k_1^2 = k_2^2 = m_1^2 = 2\lambda\rho_0^2, \qquad k_3^2 = k_4^2 = 0. \tag{A.59}$$

Conservation of momentum then implies that

$$\begin{aligned} 0 &= k_1 + k_2 + k_3 + k_4, & k_1 \cdot k_2 &= k_3 \cdot k_4 - m_1^2, \\ k_1 \cdot k_3 &= k_2 \cdot k_4, & k_1 \cdot k_4 &= k_2 \cdot k_3. \end{aligned} \tag{A.60}$$

Employing these relations, we see that, on shell, all six orderings are equal, with

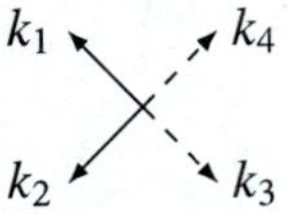

$\rho\rho\sigma\sigma$
$=$
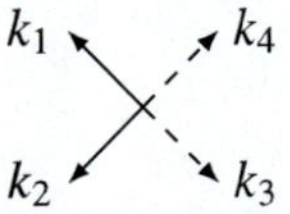

$\sigma\sigma\rho\rho$

$=$
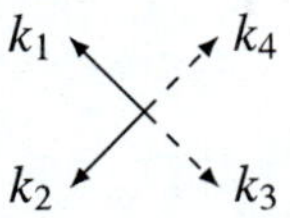

$\rho\sigma\rho\sigma$
$\rho\sigma\sigma\rho$
$=$
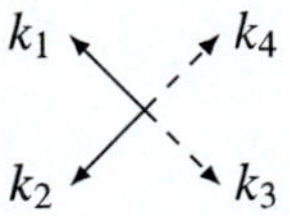

$\sigma\rho\rho\sigma$
$\sigma\rho\sigma\rho$

$$= 4i\lambda - 4i\frac{k_3 \cdot k_4}{\rho_0^2}. \tag{A.61}$$

Notice that the expression (A.61) is also invariant under $k_1 \leftrightarrow k_2$ and $k_3 \leftrightarrow k_4$.

Thus, the ordering of this vertex does not matter when the particles are on shell. However, any quantum calculation will involve off shell particles and for these, the ordering will make a difference. As discussed in Sect. 2.5.1, the correct approach is to symmetrise over all possible orderings and, thus, use the fully symmetrised rule

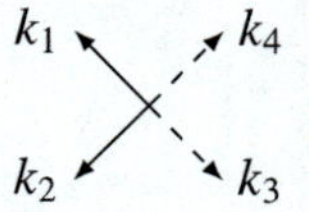

$$= \frac{2i\lambda}{3} - \frac{2i}{3}\frac{(k_1 - k_3)\cdot(k_2 - k_4) + (k_1 - k_4)\cdot(k_2 - k_3)}{\rho_0^2}. \quad \text{(A.62)}$$

Let us now calculate the $\rho\rho$ self-energy. We find

$i\Gamma_{\rho\rho}(p) =$

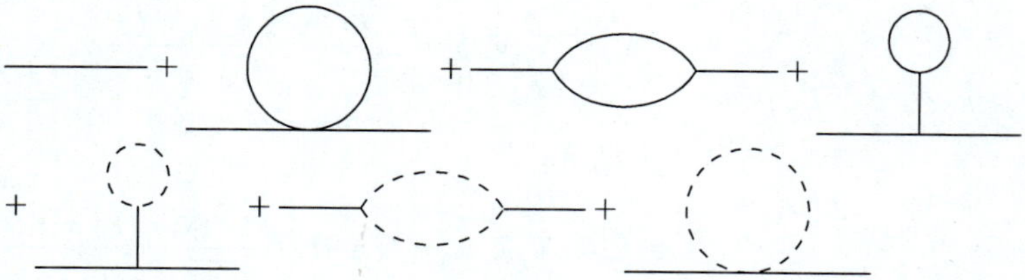

$$= i(p^2 - m_1^2) + \frac{3i\lambda}{(4\pi)^2}A(m_1^2) + 18i\frac{\lambda^2\rho_0^2}{(4\pi)^2}B_0(p^2, m_1, m_1) - 18i\frac{\lambda^2\rho_0^2}{(4\pi)^2 m_1^2}A(m_1^2)$$
$$- 12i\frac{\lambda^2\rho_0^2}{(4\pi)^2 m_1^2}A(0) + 8i\frac{\lambda^2\rho_0^2}{(4\pi)^2}B_0(p^2, 0, 0)$$
$$+ \frac{i}{2}\left[\frac{1}{(4\pi)^2}\left(\frac{2}{3}\lambda + \frac{p^2}{\rho_0^2}\right)A(0) + \frac{1}{\rho_0^2}\int\frac{d^4k}{(2\pi)^4}\right]$$
$$= i(p^2 - m_1^2) + \frac{i\lambda m_1^2}{(4\pi)^2}\left[7C_{\text{UV}} + 6\ln\left(\frac{m_1^2}{\mu^2}\right) + 4\ln\left(\frac{p^2}{\mu^2}\right) - 2\right.$$
$$\left. + 9\int_0^1 dx \ln\left(\frac{m_1^2 - x(1-x)p^2 +}{\mu^2}\right)\right]. \quad \text{(A.63)}$$

Notice that in the covariant approach, this expression is fully renormalisable, in contrast to the result obtained using the standard approach. In order to compare to the previous calculation, we set the Higgs field on shell and, hence, calculate

$$\Gamma_{\rho\rho}(p^2 = m_1^2) = \frac{\lambda m_1^2}{(4\pi)^2}\left[7C_{\text{UV}} - 7\ln\left(\frac{m_1^2}{\mu^2}\right) + 20 - 3\sqrt{3}\pi\right]. \quad \text{(A.64)}$$

This is in agreement with (A.52).

Next we calculate the self-energy of the σ field, which is given by

$$i\Gamma_{\sigma\sigma}(p) =$$

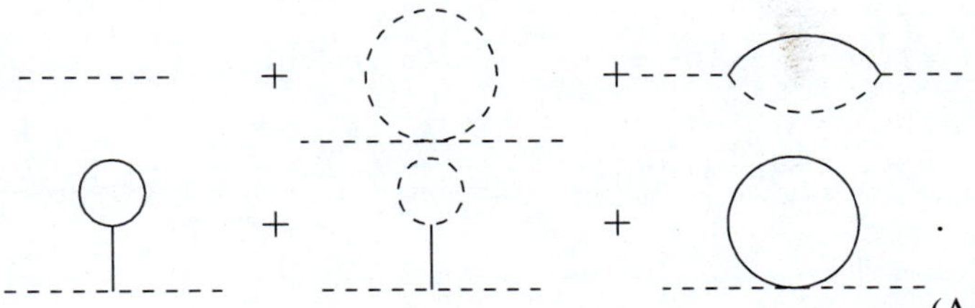

(A.65)

Before completing this calculation, let us focus on the last diagram, for which the ordering of the Feynman rule will be important. Although we will eventually symmetrise over all possible orderings, let us first consider them all individually. We find

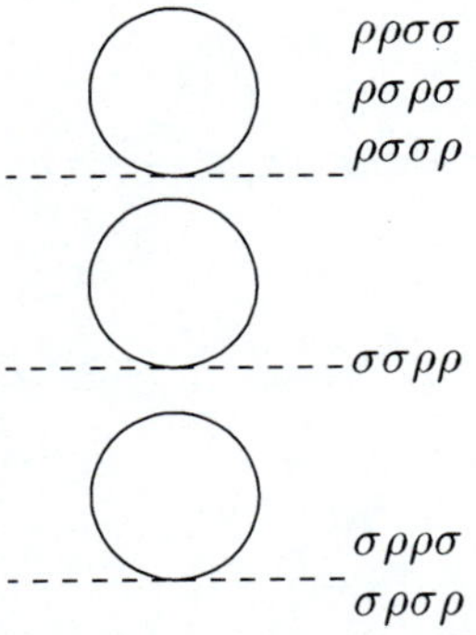

$$= -\frac{2i\lambda}{(4\pi)^2} A(m_1^2),$$

$$= -\frac{2i\lambda}{(4\pi)^2} \left(1 + \frac{1}{2}\frac{p^2}{m_1^2}\right) A(m_1^2), \quad \text{(A.66)}$$

$$= -\frac{2i\lambda}{(4\pi)^2} \left(1 + \frac{1}{4}\frac{p^2}{m_1^2}\right) A(m_1^2).$$

As expected, the ordering of the Feynman rule affects the result of this diagram. However, we note that when the external particles are on shell (i.e. $p^2 = 0$) the results converge to a single expression. This is a separate observation to (A.61) since the particle in the loop is still off shell in these calculations.

When considering external particles with off shell momentum, we must use the symmetrised Feynman rule, and, thus, we have

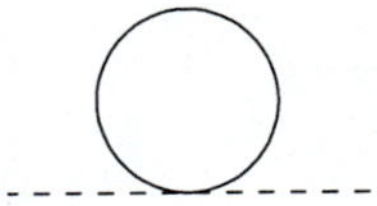

$$= -\frac{2i}{(4\pi)^2} \lambda \left(1 + \frac{1}{6}\frac{p^2}{m_1^2}\right) A(m_1^2). \quad \text{(A.67)}$$

In this way, we find

$$
\begin{aligned}
\Gamma_{\sigma\sigma} &= p^2 + \frac{12\lambda}{(4\pi)^2}A(0) + 16\frac{\lambda^2\rho_0^2}{(4\pi)^2}B_0(p^2, m_1^2, 0) \\
&\quad - 12\frac{\lambda^2\rho_0^2}{(4\pi)^2 m_1^2}A(m_1^2) - 8\frac{\lambda^2\rho_0^2}{(4\pi)^2 m_1^2}A(0) - \frac{2\lambda}{(4\pi)^2}\left(1 + \frac{1}{6}\frac{p^2}{m_1^2}\right)A(m_1^2) \\
&= p^2\left(1 - \frac{\lambda}{3}C_{\mathrm{UV}}\right) \qquad \text{(A.68)} \\
&\quad + \frac{8\lambda m_1^2}{(4\pi)^2}\left[\ln\left(\frac{m_1^2}{\mu^2}\right) - \int_0^1 dx \ln\left(\frac{m_1^2 - xp^2}{\mu^2}\right) + \frac{1}{24}\frac{p^2}{m_1^2}\left(\ln\left(\frac{m_1^2}{\mu^2}\right) - 1\right)\right].
\end{aligned}
$$

Setting the Goldstone boson on shell, we find

$$
\Gamma_{\sigma\sigma}(p^2 = 0) = 0 \tag{A.69}
$$

in agreement with (A.54).

Finally, let us calculate the tree-level S matrix element for $\rho\rho \to \sigma\sigma$ in the covariant approach. The contributing diagrams are

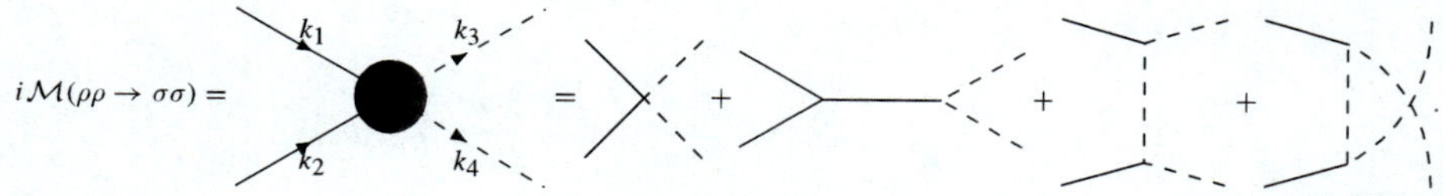

(A.70)

As discussed earlier, the ordering in the first diagram does not matter when all particles are on shell. Therefore, we have

$$
\begin{aligned}
\mathcal{M}(\rho\rho \to \sigma\sigma) &= 2\left(2\lambda - \frac{s}{\rho_0^2}\right) - 24\lambda^2\rho_0^2\frac{1}{s - m_1^2} - 16\lambda^2\rho_0^2\frac{1}{t} - 16\lambda^2\rho_0^2\frac{1}{u} \\
&= -\frac{2}{\rho_0^2}\left[3\frac{s^2}{s - m_1^2} + 2\frac{(m_1^2 - t)^2}{t} + 2\frac{(m_1^2 - u)^2}{u}\right].
\end{aligned} \tag{A.71}
$$

Notice that this result coincides with (A.56).

A.3 Example with Hidden Interactions

As a final example, let us consider a theory with Lagrangian

$$
\mathcal{L} = \frac{1}{2}\partial_\mu\rho\partial^\mu\rho + \frac{1}{2}\left(\frac{\rho}{\rho_0}\right)^2\partial_\mu\sigma\partial^\mu\sigma - t_\rho\rho - \frac{1}{2}m^2\rho^2. \tag{A.72}
$$

We see that this is a theory with a flat field space and no interactions between the ρ and σ fields in the potential. Nevertheless, the theory described by (A.72) *is* interacting as we shall see.

The vacuum for this theory is

$$\langle\rho\rangle = \rho_0 \equiv -t_\rho/m^2, \qquad \langle\sigma\rangle = 0. \tag{A.73}$$

We can, therefore, consider perturbations about this vacuum by defining

$$\rho = \rho_0 + \delta\rho, \qquad \sigma = 0 + \delta\sigma \tag{A.74}$$

as before.

Let us calculate the covariant Feynman rules for this theory. Denoting $\delta\rho$ by a solid line and $\delta\sigma$ by a dashed line, we find the following covariant Feynman rules:

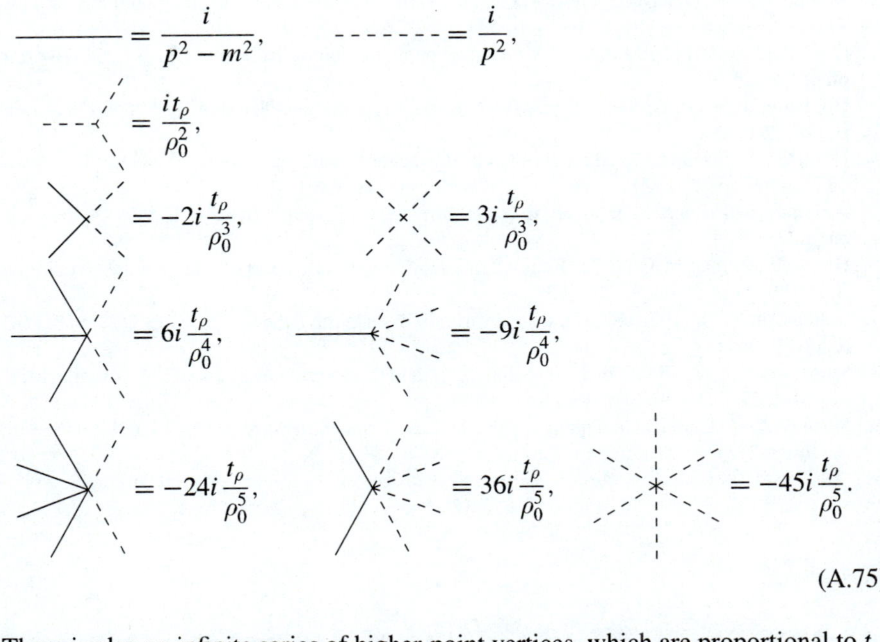

$$= \frac{i}{p^2 - m^2}, \qquad = \frac{i}{p^2},$$

$$= \frac{it_\rho}{\rho_0^2},$$

$$= -2i\frac{t_\rho}{\rho_0^3}, \qquad = 3i\frac{t_\rho}{\rho_0^3},$$

$$= 6i\frac{t_\rho}{\rho_0^4}, \qquad = -9i\frac{t_\rho}{\rho_0^4},$$

$$= -24i\frac{t_\rho}{\rho_0^5}, \qquad = 36i\frac{t_\rho}{\rho_0^5}, \qquad = -45i\frac{t_\rho}{\rho_0^5}. \tag{A.75}$$

There is also an infinite series of higher-point vertices, which are proportional to t_ρ.

We see that, when calculated using covariant techniques, this theory is revealed to have a rich structure of interactions. This includes interactions between $N > 4$ particles that would not have been present if we had derived the Feynman rules in the usual way.

To better understand why these interactions exist, we switch to a canonical parametrisation and, thus, define

$$\phi_1 = \rho\cos\left(\frac{\sigma}{\rho_0}\right), \qquad \phi_2 = \rho\sin\left(\frac{\sigma}{\rho_0}\right). \tag{A.76}$$

We then see that (A.72) takes the form

$$\mathcal{L} = \frac{1}{2}\partial_\mu\phi_1\partial^\mu\phi_1 + \frac{1}{2}\partial_\mu\phi_2\partial^\mu\phi_2 - \frac{1}{2}m^2(\phi_1^2 + \phi_2^2) - t_\rho\sqrt{\phi_1^2 + \phi_2^2}. \tag{A.77}$$

With this parametrisation, the field space metric becomes manifestly Euclidean and, thus, ordinary and covariant Feynman rules will be identical.

We see that the source of the interactions is the final term, which is non-polynomial. Thus, when this term is Taylor expanded it will lead to an infinite series of interactions, which is precisely what we discovered using the covariant technique.

References

1. Nambu Y (1960) Quasiparticles and Gauge invariance in the theory of superconductivity. Phys Rev 117:648–663. Taylor JC (ed) (cit. on pp. 26, 262)
2. Goldstone J (1961) Field theories with superconductor solutions. Nuovo Cim 19:154–164 (cit. on pp. 26, 262)
3. Goldstone J, Salam A, Weinberg S (1962) Broken symmetries. Phys Rev 127:965–970 (cit. on pp. 26, 262)
4. Hooft G (1973) Dimensional regularization and the renormalization group. Nucl Phys B 61:455–468 (cit. on pp. 27, 263)
5. Weinberg S (1973) New approach to the renormalization group. Phys Rev D 8:3497–3509 (cit. on pp. 27, 263)
6. Hooft G, Veltman MJG (1972) Regularization and renormalization of Gauge fields. Nucl Phys B 44:189–213 (cit. on pp. 27, 263)
7. Callan Curtis G Jr (1970) Broken scale invariance in scalar field theory. Phys Rev D 2:1541–1547 (cit. on p. 265)
8. Symanzik K (1970) Small distance behavior in field theory and power counting. Commun Math Phys 18:227–246 (cit. on p. 265)
9. Sbisà F (2015) Classical and quantum ghosts. Eur J Phys 36:015009. arXiv: 1406.4550 [hep-th] (cit. on p. 273)